MULTI-OBJECTIVE FOREST PLANNING

Managing Forest Ecosystems

Volume 6

Series Editors:

Klaus von Gadow

Georg-August-University,
Göttingen, Germany

Timo Pukkala

University of Joensuu,
Joensuu, Finland

and

Margarida Tomé

Instituto Superior de Agronomía,
Lisbon, Portugal

Aims & Scope:

Well-managed forests and woodlands are a renewable resource, producing essential raw material with minimum waste and energy use. Rich in habitat and species diversity, forests may contribute to increased ecosystem stability. They can absorb the effects of unwanted deposition and other disturbances and protect neighbouring ecosystems by maintaining stable nutrient and energy cycles and by preventing soil degradation and erosion. They provide much-needed recreation and their continued existence contributes to stabilizing rural communities.

Forests are managed for timber production and species, habitat and process conservation. A subtle shift from *multiple-use management* to *ecosystems management* is being observed and the new ecological perspective of *multi-functional forest management* is based on the principles of ecosystem diversity, stability and elasticity, and the dynamic equilibrium of primary and secondary production.

Making full use of new technology is one of the challenges facing forest management today. Resource information must be obtained with a limited budget. This requires better timing of resource assessment activities and improved use of multiple data sources. Sound ecosystems management, like any other management activity, relies on effective forecasting and operational control.

The aim of the book series ***Managing Forest Ecosystems*** is to present state-of-the-art research results relating to the practice of forest management. Contributions are solicited from prominent authors. Each reference book, monograph or proceedings volume will be focused to deal with a specific context. Typical issues of the series are: resource assessment techniques, evaluating sustainability for even-aged and uneven-aged forests, multi-objective management, predicting forest development, optimizing forest management, biodiversity management and monitoring, risk assessment and economic analysis.

Multi-objective Forest Planning

Edited by

Timo Pukkala
University of Joensuu,
Finland

KLUWER ACADEMIC PUBLISHERS
DORDRECHT / BOSTON / LONDON

A C.I.P. Catalogue record for this book is available from the Library of Congress.

ISBN 1-4020-1097-4

Published by Kluwer Academic Publishers,
P.O. Box 17, 3300 AA Dordrecht, The Netherlands.

Sold and distributed in North, Central and South America
by Kluwer Academic Publishers,
101 Philip Drive, Norwell, MA 02061, U.S.A.

In all other countries, sold and distributed
by Kluwer Academic Publishers,
P.O. Box 322, 3300 AH Dordrecht, The Netherlands.

Printed on acid-free paper

Printed in the Netherlands.

FOREWORD

The traditional use of forests is multi-objective. Forests have been used for hunting, as a source of firewood and construction materials and for collecting various non-wood forest products such as rattan, resin, berries and mushrooms. Forests also protect soil from erosion, regulate water resources and provide various amenities. Some forests also have cultural and religious significance.

This traditional mode of forest use has at least temporarily changed into almost pure timber production. Forests have been managed to maximise timber production or economic profit, completely neglecting the other forest uses. Fortunately, this trend has been rather temporary in most cases, and the direction of forestry has been gradually reversed or re-directed. In several countries, multiple-use forest management has already replaced single-purpose timber production for some decades. Multiple-use forestry considers activities like recreation and hunting together with timber production. Recently, ecologically sustainable forestry has gained acceptance. The target of ecological planning is to manage the forest in such a way that the populations of species remain viable and the biodiversity of the forest is maintained.

The scientific methods devised for forest planning support timber production. Many planning methods completely ignore the existence of forest functions other than wood production. The realisation that the forest planning methods available today do not correspond to the needs of today's forestry has activated forest researchers to develop and adopt new methodologies and approaches, which are specifically aimed at multi-objective situations.

This book is about the quantitative approach to multi-objective forest planning. The emphasis is on topics that are rather new and not yet systematically applied in forest planning practice. By reading the book, a planning specialist, student or a researcher will get an insight into some of the current developments in forest planning research.

The book consists of eight chapters. Chapter 1 is a general introduction to multi-objective forest planning, and especially to those aspects of planning which are emphasised in this book. Although the presentation is rather succinct, Chapter 1 does not necessarily require previous knowledge on forest planning.

Chapters 2 to 5 deal with decision analysis. They discuss methods that can be used to evaluate and rank decision alternatives in multi-objective planning situations. Chapter 2 provides theoretical foundations to the measurement of preferences while Chapter 3 is a review of complete methodologies available for multiple criteria comparison. Chapter 4 discusses how one of the most important forest management objectives, sustainability, can be evaluated in a systematic way in current multi-party and multi-objective forest planning.

The use of GIS in forest planning has gained importance due to the increased use of ecological objectives and spatial considerations. GIS can be used to produce adjacency and other spatial information for the calculation of spatial objectives and constraining variables for planning models. GIS can also be used in other pre-optimisation analyses, for instance to zone the forest into sub-areas that deserve different attention in planning. Several post-optimisation GIS analyses can also be conducted to test and evaluate candidate plans. Chapter 5 shows that GIS can also support decision analysis in a more direct way. The chapter gives an example of the use of GIS for settling forestland allocation disputes in a multi-party planning problem.

Chapters 6 and 7 deal with optimisation, especially with heuristics and spatial optimisation. Chapter 6 gives an overview of the use of heuristics in forest planning problems and the design of planning models in multi-objective spatial problems. Chapter 7 reviews recent Swedish research that has been conducted to integrate ecological spatial objectives into numerical forest planning.

Chapter 8 deals with modelling and calculation issues. The chapter reviews the recent Finnish research on the modelling of non-wood forest products and ecological variables. The chapter discusses various ways to include management objectives that are related to non-wood forest products and ecological processes into numerical forest planning. The chapter gives examples on the use of ecological and multiple-use goals in tactical planning. The example plans have been compiled using optimisation techniques and planning models similar to those described in chapters 6 and 7.

I wish to thank all the authors of this book for their valuable contributions. I also thank Ljusk-Ola Eriksson, Klaus von Gadow, Annika Kangas, Jyrki Kangas, Mikko Kurttila, Gil Mendoza and Tuula Nuutinen for helping me to review the articles included in this book. Finally, I thank Susan Kauffman for revising the English of the book.

Timo Pukkala

CONTENTS

Chapter 1

INTRODUCTION TO MULTI-OBJECTIVE FOREST PLANNING

TIMO PUKKALA
University of Joensuu, Faculty of Forestry, P. O. Box 111, 80101 Joensuu, Finland

Abstract. The purpose of forest planning is to support forestry decision-making by suggesting management alternatives, providing information about their consequences, and helping the decision maker to rank the alternatives. In multi-objective forest planning, forest plans are evaluated using various multiple criteria decision support methods and multi-objective optimisation algorithms. Multiple criteria comparison methods help to systematise subjective evaluations whereas multi-objective optimisation seeks the best plan among a huge number of alternatives using automated computer-based search methods. The ranking of alternatives depends on the preferences of the decision maker, both in multiple criteria comparison and in multi-objective optimisation. A careful analysis of preferences is an important step of any multi-objective planning case. The quantitative approach to decision-making suggests that a specific planning model be developed for every planning situation. This model is then solved, the result being a candidate plan that must pass various post-optimisation tests and analyses. There are several ways to prepare a multi-objective planning model, based on linear programming, goal programming, penalty functions or multi-attribute utility theory. The planning model may be solved using mathematical programming techniques or various heuristics. The use of heuristic optimisation has gained popularity in forest planning along the increasing importance of ecological forest management goals, which are often described with spatial variables. Examples of heuristics available to multi-objective forest planning are random ascent heuristics, simulated annealing, tabu search and genetic algorithm. Practical forest plans are produced in a computerised system, which includes sub-systems for data management, simulation of stand development, planning model generation and optimisation, and subjective evaluation of alternative plans.

Key words: decision support system, planning system, utility function, optimisation, heuristics.

Timo Pukkala (ed.), Multi-objective forest planning, 1–19.

1. PLANNING AND DECISION ANALYSIS

1.1 What is planning?

The purpose of production planning is to find the optimal allocation of production factors. The task of forest planning is to show the way to use forest resources in such a way that the welfare, or utility, of the forest owner is maximised. This definition encompasses two important aspects of planning. First, planning is always optimisation by nature, even when numerical optimisation methods are not used. Secondly, planning aims at maximising the benefit, or utility, of forest owner. To be able to maximise utility, the planner should know which forest products and features bring utility to the owner. For this, forest planning needs to estimate the preferences of the owner.

Practical forest planning is not only preference analysis and optimisation. Inventory, data management, calculation and computer simulation all play important roles in the process that leads to a forest plan. Forest planning compares alternative forest management plans, which may also be called decision alternatives, aiming at finding a decision that has the best possible consequences. Various models and calculation techniques are used to predict the consequences of decision alternatives. This requires predicting the stand development under alternative management options and calculating the amounts of those products and services which are important to the owner. The predictions use computer simulation and are based on forest inventory data stored in a database as well as on models and calculation methods.

It is not possible to tell the starting date of multi-objective planning. It is not even easy to say when forest planning is single-objective and when it is multi-objective. Forest planning that maximises the net present value without any other goals may also be considered as multi-objective planning; the net incomes from different time periods, the discounting factors indicating their importance, or the time preferences of the forest owner, can be interpreted as different objectives. Linear programming (LP) with a single objective variable can also be seen as a multi-objective planning method: one objective is handled through the objective function while the others, such as the even-flow objective, are taken care of by the constraints of the LP model.

However, what is clear is that forest planning is more clearly multi-objective than before, and objectives representing forest uses other than timber production have become more common. In several countries, timber management planning was replaced by multiple-use planning already some decades ago. Multiple-use planning considers activities like recreation and hunting together with timber production. Recently, ecological planning has gained importance. The task of ecological planning is to find such a management for the forest that the populations of plant and animal species remain viable, i.e., the biodiversity of the forest is maintained. Ecological planning does not usually mean that timber production and multiple-use objectives no longer exist; ecological goals are just additional objective variables making the planning more multi-objective than earlier and, unfortunately, more complicated.

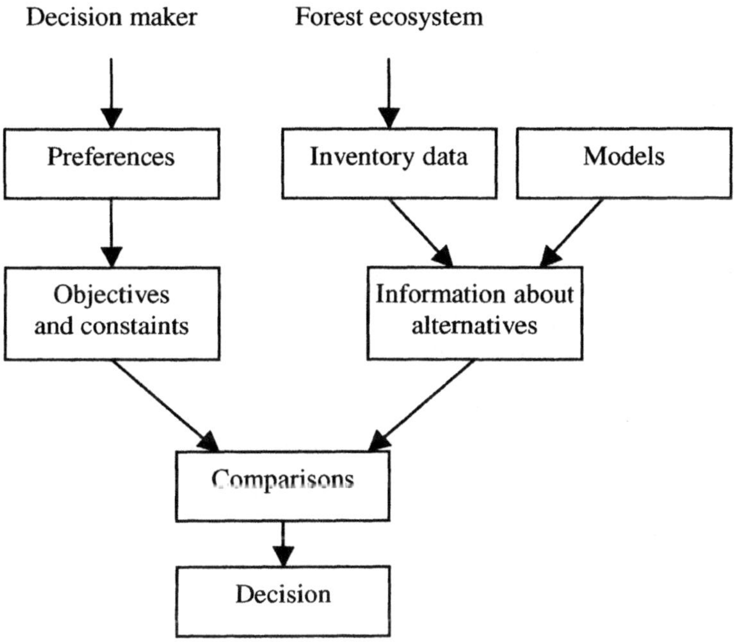

Figure 1. The elements of forestry decision-making. All steps that precede Decision *belong to decision analysis. Planning is a part of decision analysis.*

Timber management planning serves the economic function of forest whereas ecological planning maintains its ecological functions. Multiple-use planning is also related to the forest's social functions because the multiple uses are typically customary forest uses, and one aspect of social sustainability is to respect and maintain these uses. Another element of social sustainability is the fairness and transparency of the process in which forest plans are produced and management alternatives evaluated. This aspect of social sustainability calls for participatory planning which involves several stakeholders, or interest groups, with their own and usually conflicting goals and expectations. Participatory planning certainly increases the number of issues and objectives to be considered in forest planning, as compared to planning for a single decision maker.

1.2 What is decision analysis?

Good forest planning supports the decision making of a forest owner. This support consists of producing decision alternatives (alternative forest plans) to choose among, predicting their consequences, and ranking the alternatives in a systematic way (Fig. 1). The ranking of alternatives is always based on subjective preferences. Despite this subjectivity the ranking process can be automated using numerical optimisation algorithms. If numerical optimisation is not used, it is possible to apply various comparison techniques such as the Analytic Hierarchy Process (Saaty 1980) to weigh decision criteria and rate decision alternatives. These techniques are examples of decision analysis tools.

If the alternatives are ranked by numerical optimisation, the preferences of the forest owner and other stakeholders must be learned before the optimisation step. Otherwise, it is not known how to formulate the objective function and the planning model, the solution of which is a plan that maximises the utility of the owner. Often, the preferences are found out in discussion with the owner and other relevant parties. However, there are also more sophisticated and systematic ways to measure preferences.

Forest planning is often seen as an activity that deals only with the objective steps of the decision-making process (Fig. 1). According to this view, planning includes producing decision alternatives, predicting their consequences, and ranking the alternatives with numerical optimisation. Objective planning combined with subjective considerations is called decision analysis. Decision analysis is a wider concept than planning: it covers planning as well as the analysis of preferences and subjective valuation of alternatives. It is decision analysis rather than planning which supports modern forestry decision-making. Therefore, forest planners should expand their knowledge to cover not only planning tools, but also decision analysis methodologies.

1.3 Strategic and tactical planning

Forest planning may be classified as strategic, tactical and operative (Fig. 2). Strategic planning helps in setting the main direction of forestry so that the forest and its use contribute in the best possible way to the welfare of the owner. Strategic planning tries to answer the question: "What should be the overall goal in the forest?" Strategic planning helps to specify management objectives for tactical planning. Tactical planning answers the question of what to do in the forest to meet the production targets derived in strategic planning and the consequent decision-making. Tactical planning produces a list of management actions (operations or treatments) that should be performed in the forest. Operative planning devises the implementation of the treatments prescribed in tactical planning.

Typical to strategic planning is that just a few alternative plans clearly different from each other, called strategies, are evaluated. The evaluation considers the preferences of the owner as well as the risk and uncertainty related to the outcomes of decision alternatives. Because the number of alternatives is often small and subjective considerations dominate in the evaluation, the ranking of alternatives is conveniently done using various multiple criteria comparison methods rather than numerical optimisation. Although the strategy alternatives may be produced using optimisation, these solutions are simply alternatives that need to be evaluated and compared in subsequent analyses.

In tactical planning the uncertainty is considered less relevant and the objectives more clearly defined than in strategic planning. Therefore, optimisation is commonly used to systematically scan the decision space to find the single best alternative among a huge number of alternatives. Other differences between strategic and tactical planning are that strategic planning covers a longer time horizon than tactical planning, strategic planning is not location-specific while tactical planning is that, and more of the production factors (resources) are considered variable in strategic planning than in the tactical planning.

This comparison suggests that different ranking methods are suitable for strategic and tactical planning. However, despite these differences it is not necessary that strategic planning always precedes tactical planning, and tactical planning precedes operational planning. It is possible to use the same data, time horizon and planning methods in all levels, and integrate the three levels to various degrees. If separate strategic planning is not conducted, or it is integrated with the tactical planning, which is typical with small forest holdings, the preference analysis carried out before the optimisation serves the same purpose as the separate strategic planning. Therefore, the analysis of preferences may be regarded as the most strategic step of planning.

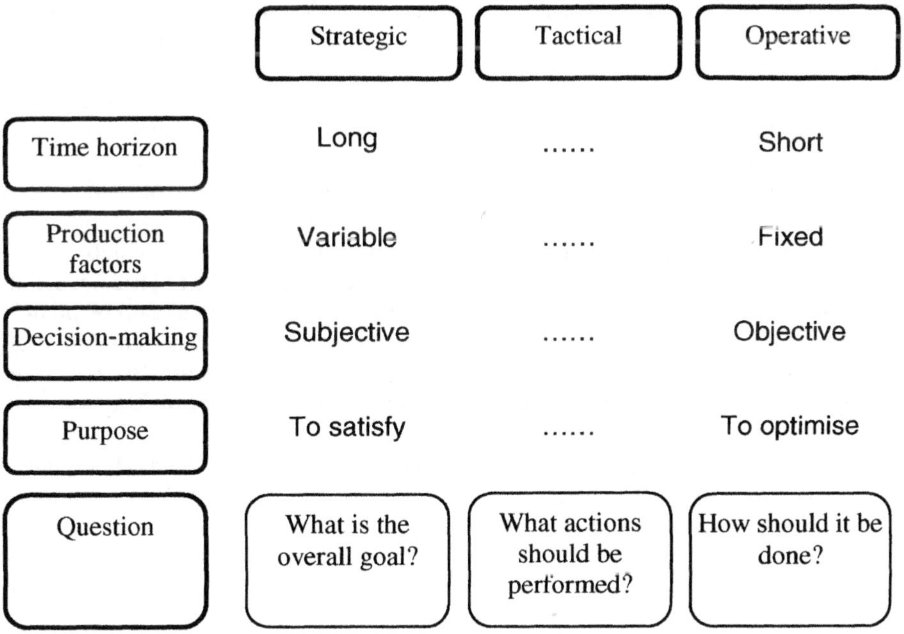

Figure 2. A comparison on strategic, tactical and operative planning.

The reason that forest planning has been split into the three levels is not to improve planning but to simplify the very complicated planning task. The most efficient plans can probably be achieved if the three planning levels are completely integrated. If operative considerations, like combining the proposed cuttings into operationally meaningful blocks, are included in tactical planning, the resulting plan is obviously "better" than without these considerations. The integration of operative aspects into tactical planning often calls for using spatial optimisation.

2. QUANTITATIVE PLANNING

2.1 Why quantitative planning?

Current forestry decision-making situations are complex for many reasons. There are many goals and conflicting issues that should be taken into account. Several parties are involved in planning, and different parties have different goals and expectations. The risk and uncertainty related to the selection of a forest plan is high for several reasons. The future states of factors like timber prices and growth variation of trees due to climate are unknown. The models used to predict stand development may contain errors and they can predict only a part of the variation in tree growth. Predicting the multiple-use outputs and ecological quality is even more uncertain due to limited empirical knowledge and modelling experience. In addition, the future preferences of decision makers, i.e. their goals, risk preferences, and time preferences are seldom if ever known with certainty, even when sophisticated preference analysis tools are used. Even the current situation of the forest is not known exactly because the information is based on sampling or visual assessments, and only a part of the site and growing stock characteristics are surveyed.

The main thing that makes forest planning complicated is the huge number of decision alternatives. Decision alternatives are all plans that can be prepared for the forest. Typically, a forest plan is produced in two steps so that the first step specifies alternative management schedules for each stand or compartment. These schedules are sequences of treatments over the planning horizon. The second step of planning seeks the best combination of treatment schedules for stands. The suitability of a plan depends on objectives specified at the level of the whole forest. Because stands are regarded as the smallest treatment units (indivisible) the idea is to choose exactly one treatment schedule for every stand.

All different combinations of stands' treatment schedules are alternative plans. If the number of treatment schedules is the same for all stands, the total number of alternative plans can be computed from

$$P = T^S \tag{1}$$

where P is the number of alternative plans, T is the number of treatment schedules per stand, and S is the number of stands. If the forest consists of just two stands and both have three treatment options, the total number of plans is $3^2 = 9$ (Table 1). If there is a third stand, the number of alternative plans is 3 x 9 = 27 because all the combinations of treatment schedules of stands 1 and 2 (Table 1) can be repeated with each of the three treatment schedules for stand 3.

Table 1. Number of alternative forest management plans in a forest consisting of two stands and both stands having three management options.

Plan number	*Treatment of stand 1*	*Treatment of stand 2*
1	1	1
2	1	2
3	1	3
4	2	1
5	2	2
6	2	3
7	3	1
8	3	2
9	3	3

In a very small forest holding consisting of 30 stands, each having three alternative treatment schedules, the total number of different plans is 3^{30} = 205,891,000,000,000. This is 41,000,000 plans for every Finn. If a Finn needs 10 seconds to evaluate one plan, it would take 14,300 working days for all Finns to evaluate all the plans.

This example shows clearly that it is not possible to compare and evaluate all decision alternatives available in forest planning without relying on efficient numerical tools. Most forest planning situations are much more complicated than the above example because there are more stands and more treatment alternatives per stand. In many practical planning situations the number of alternative plans is far too high even for computers to evaluate all plans. This reality calls for using such solution techniques that need to evaluate only a small fraction of alternatives to reach a plan that is optimal or close to it.

Even the evaluation of a single alternative may be difficult without quantitative methods because of the high number of decision makers and objectives, and the complicated nature of some objectives. The use of quantitative analyses sometimes allows better communication between the parties than qualitative evaluation, as well as systematic sensitivity analyses of the effect of various factors (preferences, timber prices) on the ranking of plans. In addition, quantitative analysis may also provide an explanation of the reasons a particular plan was regarded more highly than the others.

2.2 Planning model

The quantitative approach to decision making, as suggested by Management Science, proceeds in the steps shown in Figure 3 (see e.g., Render and Stair 1992, p. 5). This approach suggests that a unique model be developed for every planning and decision making situation. In the context of planning, the model is called a planning model. The planning model is filled with data and solved with a suitable procedure. The solving method is often a numerical optimisation algorithm. The algorithm that is used to solve the model also affects the problem formulation, i.e., the type of planning model used. For example, when linear programming (LP) is used as the solution method, the forest planning problem is formulated as an LP model. If goal programming (GP) is used, the problem is formulated as a GP model.

The solution suggested by the optimisation algorithm is only a candidate plan that should be carefully tested and analysed. If it passes the tests, and the sensitivity analyses show that the solution is not too sensitive to uncertain parameters, the solution may be recommended as a forest plan to be implemented.

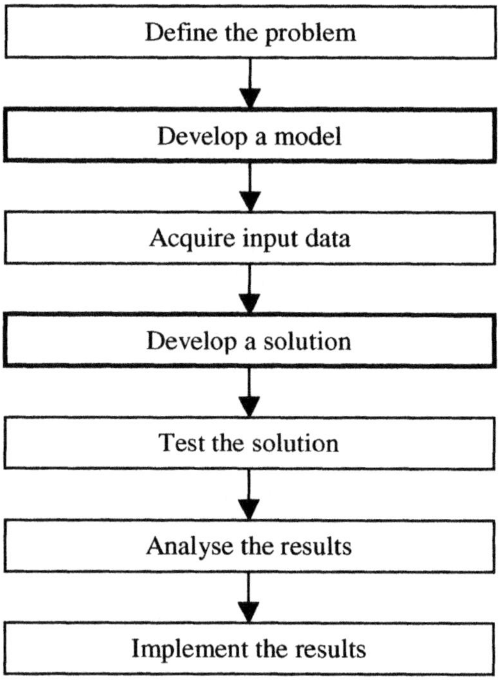

Figure 3. The quantitative analysis approach to decision making.

The data that are included in the planning model fall into two categories and come from two sources (Fig. 1):

- data on preferences, which is obtained from the decision maker (legislation reflecting society's preferences belong to this category), and

- data on forest production, which is based on the combined use of forest inventory data and models, unit prices, etc.

When LP is used to solve the planning problem, the information about preferences is accommodated in the model through the choice of the objective and constraining variables and the values of constraints (right hand side of the model). The information concerning the production possibilities of the forest is contained in the coefficients of the objective function and the constraints. The coefficients tell how much a stand treatment schedule produces or consumes the objective or constraining variable.

With linear programming the planning problem may be modelled as follows:

Maximize

$$z = \sum_{j=1}^{n} \sum_{i=1}^{n_j} c_{ij} x_{ij} \tag{2}$$

Subject to

$$\sum_{j=1}^{n} \sum_{i=1}^{n_j} a_{ijk} x_{ij} \geq B_k \qquad k = 1,\ldots, K_1 \tag{3}$$

$$\sum_{j=1}^{n} \sum_{i=1}^{n_j} a_{ijk} x_{ij} \leq B_k \qquad k = K_1+1,\ldots, K \tag{4}$$

$$\sum_{i=1}^{n_j} x_{ij} = A_j \qquad j = 1,\ldots, n \tag{5}$$

where z is objective function, n is the number of stands, n_j is the number of alternative treatment schedules in stand j, x_{ij} is the area of stand j treated according to schedule i, B_k are constraints, K is the number of constraints, and A_j is the area of stand j. Coefficient c_{ij} tells how much one hectare of stand j treated according to schedule i produces or consumes the objective variable, and coefficient a_{ijk} indicates how much one hectare of schedule i in stand j produces or consumes constraining variable k. The unknown variables, the optimal values of which are found in optimisation, are the x_{ij}, i.e., stand areas treated with different methods. Variables x_{ij} are also called decision variables, because the decision-maker can decide their values. The LP problem is usually solved using standard computer software that is based on the simplex algorithm.

The above problem formulation allows the division of a stand into two or more sub-areas that are treated differently. If this is not wanted, the problem is formulated so that the x_{ij} are specified as integer variables, and constraints (5) are converted into

$$\sum_{i=1}^{n_j} x_{ij} = 1 \qquad j = 1,\ldots, n \tag{6}$$

In this formulation, the decision variables are the proportions of compartments treated according to different schedules. This modification requires that the coefficients c_{ij} and a_{ijk} are converted from per-hectare values into per-stand values.

The LP model assumes additivity and proportionality, among other things. Additivity means that the totals of the objective and constraining variables can be computed by summing the values of the respective variables in the treatment schedules that are included in the solution. Proportionality means that objective and constraining variables are linear functions of the decision variables.

2.3 Multi-objective planning model

In multi-objective forest planning the problem is modelled in such a way that the planning model accommodates several simultaneous management objectives. There are alternative ways to accomplish this:

1 One objective is minimised or maximised via the objective function and the other goals are taken care of through constraints. This corresponds to the use of linear programming.

2 The linear programming model is modified so that the objective function measures the deviations of several objective variables from their target levels. The target levels are given in the other equations of the problem formulation. Strict constraints may be added to the problem formulation. This corresponds to the use of goal programming.

3 A single objective such as net present value or total cost appears in the objective function, which is augmented with a penalty function. The penalty function measures how much a set of additional objective variables deviate from their target values. The penalty function has the same unit as the objective variable. Constraints may be added to the problem formulation.

4 A multi-attribute utility function is developed and used as the objective function. Constraints may be added to the problem formulation.

Formulations 1 and 2 typically use mathematical programming (LP, GP) as the solution method whereas formulations 3 and 4 often rely on heuristics. However, any problem that can be solved with mathematical programming can also be solved with heuristics. Heuristics are iterative methods that first generate one or several initial solutions and then try to improve them using a predefined technique until a user-specified stopping criterion is met. Mathematical programming has the advantage that it always –when the assumptions hold– finds the optimal solution, which is not guaranteed with heuristics. The advantage of heuristics is that they can solve very complicated problems and the solution method is usually rather easy to understand.

All four planning model types listed above have an objective function and most of them also have constraints to be checked when the optimisation algorithm evaluates a candidate solution. It is possible that the same variable appears both in the objective and constraining functions, the constraint giving it a strict upper or lower bound. Also, the same variable may appear in several constraints, or it may be included in the objective function more than once.

In the LP and GP formulations all objective and constraining variables are calculated as linear combinations of decision variables (areas or proportions of compartments treated according to a particular treatment schedule). Heuristics allow non-linear functions of decision variables as well as spatial objectives and constraints, the calculation of which uses for instance the adjacency information of stands. The type of objectives and constraints, the solution method and the exact formulation of the planning model are all connected to each other. Therefore, a logical way to proceed in planning is to find out the objectives (and constraints) first, then select a suitable solution method, and finally formulate the problem so that it accommodates all the objectives and is in accordance with the solution technique.

Goal programming

When linear programming is used in multi-objective planning, one of the goal variables is selected as the objective variable while the others are controlled through constraints. However, this problem formulation does not always correspond exactly to the real-life problem because the constraints referring to the other goals may not be as strict as the problem formulation suggests. The objectives may have target levels and desirable values but it may not be absolutely necessary to reach these values.

To make the planning model correspond to the real-life situation, an alternative problem formulation, called a goal programming model, has been developed (e.g., Dykstra 1984). In this model the objective function consists of the deviations (d) of objective variables from their target values:

Minimise

$$z = \sum_{k=1}^{K} d_k^+ + \sum_{k=1}^{K} d_k^- \tag{7}$$

where d_k^+ indicates how much the target value of objective k is exceeded (surplus), and d_k^- is the quantity by which target k falls short (slack). The target levels (b_k) are given through the so-called goal constraints, one constraint per objective:

$$\sum_{j=1}^{n} \sum_{i=1}^{n_j} a_{ijk} x_{ij} + d_k^- - d_k^+ = b_k \qquad k = 1, \ldots, K \tag{8}$$

These constraints together with the objective function mean that levels b_k are pursued: deviations are regrettable (they are minimized) but they are allowable (constraints are flexible).

If only falling short of the target, but not exceeding it, is harmful, the surplus variable (d_k^+) is left out from the objective function and the corresponding goal constraint. If only exceeding is to be avoided, the slack variable (d_k^-) is dropped. It is also possible to multiply the deviation variables by constants (w_k and v_k) that reflect the relative importance of the respective objective variables (Kangas and Pukkala 1992):

Minimise

$$z = \sum_{k=1}^{K} w_k d_k^+ + \sum_{k=1}^{K} v_k d_k^- \tag{9}$$

This modification does not affect the goal constraints.

Another improvement to the model is that of scaling the deviation variables to the same range of variation (usually between 0 and 1) so that the units of measurement do not affect the actual weights of different goals. This scaling can be done by dividing the deviations by the largest possible value of the goal variable (single-objective maximum).

Minimise

$$z = \sum_{k=1}^{K} w_k \frac{d_k^+}{Q_k^{\max}} + \sum_{k=1}^{K} v_k \frac{d_k^-}{Q_k^{\max}} \tag{10}$$

Also in this case the goal constraints remain unchanged. The goal programming model can be solved with the same software and algorithms as the ordinary LP model.

Penalty functions

Another approach to multi-objective optimisation is to use penalty functions. The penalty function measures the deviations of some critical variables from their target levels. For example, an ecological planning problem may be formulated so that variables measuring the ecological quality of the forest, such as the area of old forest and volume of deadwood, have target values determined by specialists. Falling short of these targets decreases the value of the plan. In this case, the objective function may be formulated as follows:

Maximize

$$z = v \sum_{j=1}^{n} \sum_{i=1}^{n_j} c_{ij} x_{ij} - \sum_{k=1}^{K} w_k \left| S_k - T_k \right|^{\alpha} \tag{11}$$

where v is the importance (weight) of the objective variable, w_k is the importance of penalty variable k, K is the number of penalty variables, S_k the value of variable k in the current solution, and T_k is the target level of variable k. When heuristics are used to solve the problem, S_k can be any function of the decision variables x_{ij}. Parameter α tells how quickly the penalty increases as a function of deviation. It is also possible to devise the penalty function so that only exceeding the target or falling sort of it should be avoided, but not both. One more possibility is that both exceeding and falling short of the target increase penalty, but with different rates.

In addition to the objective function, the problem formulation may contain one or several constraints. A constraint stating that only one treatment schedule can be used in one stand is always required:

$$\sum_{i=1}^{n_j} x_{ij} = 1 \qquad j = 1,\ldots, n \tag{12}$$

where x_{ij} is an integer variable indicating the proportion of compartment j managed according to treatment schedule i. However, the solution algorithm may also automatically take into account the requirement of the constraint, which means that, unless there are other constraints, the explicit problem formulation requires only the objective function.

Utility functions

One more possibility in designing multi-objective planning problems is to apply multi-attribute utility theory. This theory suggests that the utility gained from the implementation of a forest plan depends on several management objectives, which may also be called decision criteria. The utility that the decision-maker will experience when implementing a plan is measured with a utility function, which may look as follows:

$$U_l = \sum_{k=1}^{K} w_k u_{kl} \tag{13}$$

where U_l is the total utility for plan l, w_k is the importance of criterion k and u_{kl} is the sub-utility that plan l produces through criterion k. The sub-utilities may be calculated on the basis of subjective evaluation, as done in multiple criteria decision support methods (see Chapter 3), or they may be computed from sub-utility functions. When the number of candidate plans to be evaluated is high, as in the case of numerical optimisation, sub-utility functions must be used. The utility function is converted into:

$$U_l = \sum_{k=1}^{K} w_k u_k(q_{kl}) \tag{14}$$

where u_k is the sub-utility function for criterion k, and q_{kl} is the quantity that plan l produces the criterion variable k. When the problem is solved with heuristics, q_{kl} can be a linear or a non-linear function of decision variables (x_{ij}). Its calculation may use, for instance, information on the adjacency of stands, as well as characteristics describing the growing stock and treatments.

The utility function of (14) assumes that the sub-utilities produced by different criteria are interchangeable in ratios indicated by the criterion weights (w_k). This means, for instance, that a low achievement of an ecological objective (even a zero value) may be compensated for by high values of economic goals. If this property of the utility function is not acceptable, the function may be converted into one of the following forms:

$$U_l = \prod_{m=1}^{M} w_m u_m(q_{ml}) \tag{15}$$

$$U_l = \left[\sum_{k=1}^{K} w_k u_k(q_{kl}) \right] \times \prod_{m=1}^{M} v_m u_m(q_{ml}) \tag{16}$$

where subscript m refers to goals that are not interchangeable with the other goals.

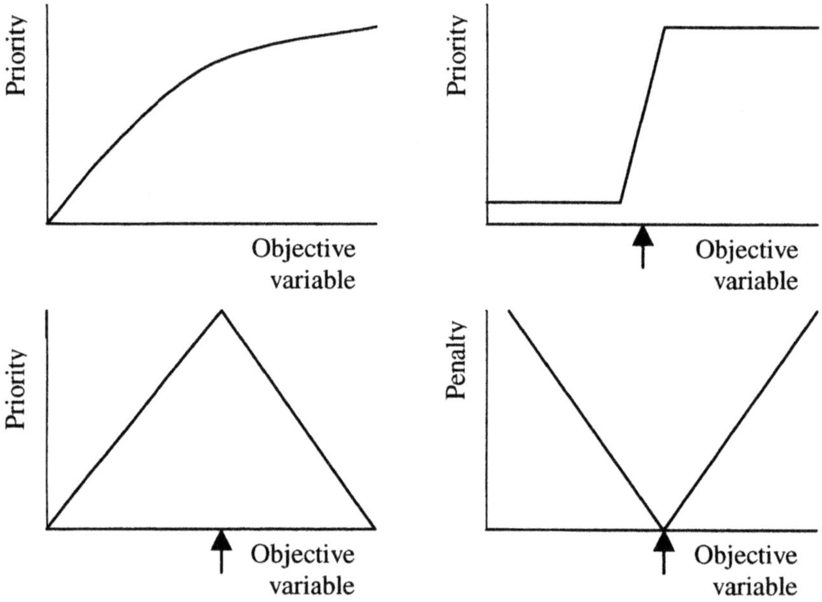

Figure 4. Examples of sub-utility functions. The top-left function represents decreasing marginal utility and the top-right function is a constraint-type function. The function at bottom left measures deviation from a target level (arrow) in the utility theoretic planning model. The bottom-right function shows how the deviation affects the objective function value in the goal programming and penalty function approaches.

An example of the application of utility function in multi-objective forest planning is the Hero method proposed by Pukkala and Kangas (1993), which has been used for a decade both in spatial (Pukkala et al. 1995) and non-spatial planning problems (e.g., Kangas et al. 1996). In this method, the problem formulation requires only the utility function with no constraints at all. The requirement that only one treatment schedule per stand can enter the solution is implicitly taken into account by the optimisation algorithm. The different nature of different objectives (objective-type, constraint-type) is taken into account by the shape of the sub-utility functions (Fig. 4).

The use of multi-attribute utility theory requires a method to estimate the preferences of the decision makers. In the utility functions shown above, the preferences refer to the importance of different management goals (w_k and v_k) and to the sub-utility functions, which indicate the priority of different amounts of goal variables. Various techniques based on direct rating or paired comparisons of objective variables, or different amounts of an objective variable, are available for estimating the preferences (see Chapters 2 and 3).

In addition to the utility-theoretic planning approach, the same preference analysis techniques can also be used with goal programming to estimate the coefficients v_k and w_k of Equation 10 or the penalty function coefficients v and w_k in Equation 11. Therefore, the goal programming, penalty function and utility function approaches can all apply utility theory.

2.4 Heuristics

The Hero method uses a simple heuristic algorithm to maximise the utility function. However, the same utility function can be maximised with several other heuristics, and the Hero algorithm can be used with problem formulations other than the utility theoretic formulation. Most heuristic methods first produce an initial solution and then gradually improve it by making local changes, so-called moves, in the solution (typically changing the treatment schedule of one stand). Some heuristics (genetic algorithms) produce several initial solutions and then combine them so that the set of solutions gradually improves. After a user-specified stopping criterion the algorithm stops, and suggests the best solution found during the process as the optimal solution to the planning problem.

The simplest heuristic first selects a random treatment schedule for every compartment. It then takes a random compartment and changes its treatment schedule. If the change improves the objective function (and satisfies the constraints) it is accepted, otherwise it is rejected. This heuristic method may be called random descent if the objective function is minimised, or random ascent if the objective function is maximised. The Hero algorithm highly resembles the random ascent method, the difference being that Hero inspects stands and their treatment schedules sequentially rather than randomly.

A heuristic called simulated annealing also resembles the random ascent method (e.g., Reeves 1993). The difference is that simulated annealing also accepts inferior solutions, i.e., changes that worsen the objective function value. Improving moves are always accepted and inferior solutions are accepted with a probability much lower than one. This probability decreases during optimisation with a rate determined by the user-specified "cooling schedule" of the process.

Another heuristic method is a tabu search. It also begins with an initial solution which is then gradually improved. The specific feature of tabu search is that recent changes are remembered, and repeating them is prohibited (made tabu) for some time. The idea is to force the algorithm to find new combinations of stand treatment schedules. Forbidding the repetition of recent changes decreases the probability that the process gets stuck to a local optimum.

Genetic algorithms require several initial solutions, which in this method may be called "chromosomes". Pairs of chromosomes are combined to produce new ones, using a "crossing-over" technique. An example "re-combination" is taking the treatments for stands 1 to 10 from initial solution 1, and the treatments for the other stands from initial solution 2. Crossing over produces a new solution (chromosome), which may be further modified by changing the treatment schedule of a single stand ("mutation"). The new solution produced by crossing-over and mutation is evaluated using the objective function or some other "fitness function". Good chromosomes are kept alive and bad ones are discarded. Repeating this process over many "generations" results in a "population" of solutions, the average "fitness" of which is good. The best chromosome of the last generation is the proposed solution to the planning problem.

It is possible to combine and modify the heuristics described above in several ways (e.g., Boston and Bettinger 1999, Falcão and Borges 2002). It is even possible to combine mathematical programming and heuristics (e.g., Boston and Bettinger 2001, Öhman and Eriksson 2002). In a planning problem requiring spatial considerations it is possible to solve the spatial sub-problem with heuristics. The solution of the spatial problem is then used to constraint a non-spatial problem that is solved with linear programming. Another possibility is to solve a non-spatial problem first, using linear programming, and utilise the reduced costs of treatment schedules, provided by the LP method, in a penalty function of a spatial heuristic planning model. These penalties lead to favouring solutions that are close to the LP solution.

The use of heuristics in multi-objective forest planning has become common during the past decade. The popularity is related to the fact that forest planning requires spatial considerations more often than before, due especially to ecological objectives, and heuristics deal with spatial optimisation problems more ably than mathematical programming. However, good forest planning has almost always used some kind of heuristics, either as the main method, or combined with other techniques. Model manipulation, which is one kind of heuristics, is commonly used with mathematical programming. In model manipulation, modifications are made in the planning model (LP model or GP model), and the model is solved after every change. The solution (and problem formulation) that is the most satisfying to the decision-maker is finally selected. Model manipulation resembles sensitivity analysis, which is a necessary step of good numerical forest planning.

Quite often tactical forest plans are compiled without using any numerical optimisation methods. A typical procedure is to prescribe a single treatment schedule for every stand, based on silvicultural recommendations and other considerations. The forest-level outcome of these prescriptions is then calculated and evaluated. If the outcome is not regarded as acceptable (there may be too many or too few cuttings), the plan is modified. The treatment proposals of individual stands are altered until the result is satisfactory at the forest level. This procedure is also an example of heuristics, in the category of local improvement. The main difference between this procedure and the random ascent heuristic is that the process is not automated.

2.5 Planning system and decision support system

Practical plans are prepared using computerised planning systems, which may also be called decision support systems (DSS) if the emphasis in its design is on decision support. Decision support systems can be defined as computer systems that support rather than replace a decision maker (Render and Stair 1992). If only the planning system is computerised, the system may be called a planning system. Planning systems also aid in decision making, but they lack components and features that make them true decision support systems.

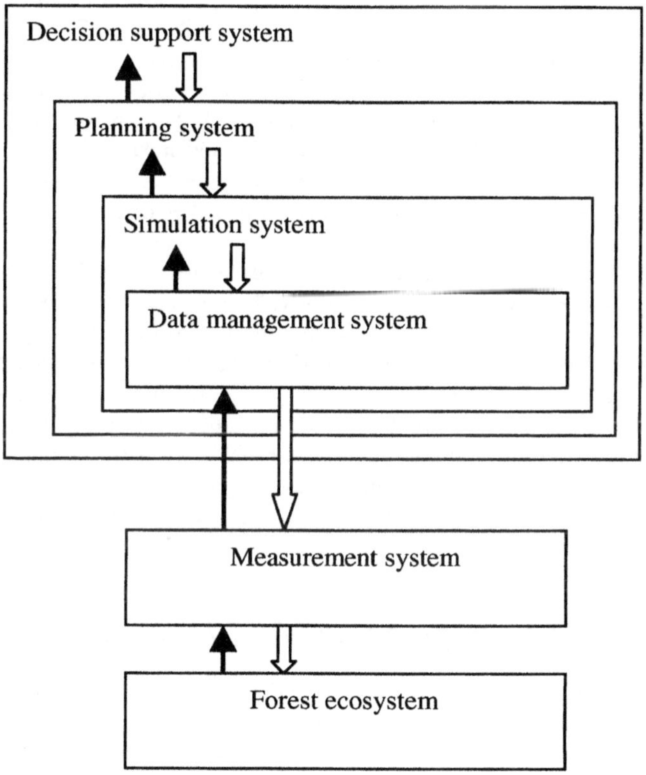

Figure 5. A computerised forestry decision support system may be seen as a sequence of sub-systems or as a set of extensions to a data management system. Black arrows represent flow of information, and open arrows represent request for information.

Figure 5 shows that a DSS is composed of different sub-systems, which may correspond to separate modules of the computer system. Forest inventory data are input to a data management system (or database management system, DBMS), which maintains the data and provides information for a calculation and simulation system. The DBMS augmented with a simulator supports decision making better than DBMS alone does because it can answer what–if questions concerning the long-term consequences of decision alternatives. A planning system includes a planning model generator and at least one solver (optimisation routine) as an additional component. A true DSS contains additional tools for estimating preferences or evaluating decision alternatives in user-friendly and subjective ways. The additional computer modules may consist of multiple criteria comparison aids (see Chapter 3), but also of forest visualisation tools which convert the information produced by the other sub-systems into a form that is easily understandable to the decision maker.

The DBMS, simulation system, planning system and a true DSS can all be called decision support systems because all of them can be used to support decision-making. In fact, the sequence of systems from DBMS to DSS also reflects the history of the development of computerised forestry decision support systems. The data management systems existed first. They were then complemented with sophisticated calculation systems and simulators. Later on the simulations of alternative stand management schedules were collected into a planning model that was solved with suitable software. The current trend is to enhance the forest planning systems so that they become true decision-maker-oriented decision support systems.

The sub-systems shown in Figure 5 indicate the categories of skills that today's forest planning expert should possess and study. The expert should know about data modelling and database management. He/she should know how to predict future stand development and how to calculate various stand- and forest-level variables, which can be used as coefficients, objectives and constraints of planning models. The planner should also be aware of alternative optimisation methodologies and their suitability to different planning problems. Finally, a skilful planner also knows techniques that can be used to systematise and support subjective comparisons of decision alternatives and to estimate the preferences of the stakeholders involved in planning.

3. NOT ONLY QUANTITATIVE PLANNING

Quantitative forest planning relies on quantitative variables and uses sophisticated optimisation algorithms and selection techniques. However, the solutions proposed by these tools are seldom if ever truly optimal for several reasons. One reason is the high degree of uncertainty related to the future and to the present. The criteria assumed in the ranking of alternatives and the way they are put to the planning model may not correspond exactly to the forest owner's true preferences, which are often non-numerical, vague and changing with time. The quantitative variables that are available in planning calculations may be insufficient for describing forest features important to the owner, such as biodiversity, hunting possibilities, or scenic beauty of the landscape.

Planning should provide support for decision making. The decision support provided by planning is not helpful if the solutions found by planning are not actually optimal, and if the planning model does not accommodate the real objectives of the decision maker. The decision support provided by planning should not only provide optimisation, but also help to evaluate the production possibilities of the forests, reveal the trade-offs between different objectives, and generate decision alternatives and predict their consequences. The decision support also comprises qualitative evaluation aids.

Several aspects of decision support can be achieved only when the decision-maker himself/herself uses the system. However, this is only possible with a simple planning system. The use and development of overly-sophisticated planning models and computer system has the danger that planning becomes specialist-centred, with a consequence that the preferences and needs of the decision-maker are forgotten, or the decision-maker is not committed to the plan.

Good decision support requires a simple system that assists subjective evaluations. On the other hand, the complexity of today's planning problem calls for efficient and objective numerical tools. How can these conflicting demands be met? The user interface to the planning system evidently plays an important role here. The interface of the system should be simple although the calculations may be complex. Therefore, simple decision support should not disregard sophisticated planning tools. Because planning has become more complicated than during the era of maximal timber production, the need for efficient tools is even more urgent now than before. Although the quantitative tools may not be able to show the true optimum, they are able to show the effects of various assumptions on the optimal management and the outcome of the plans. User-friendly decision support requires that the unfriendly forest planning tools be augmented with techniques and interfaces that convert objective planning into subjective decision support.

REFERENCES

Boston, K. and Bettinger, P. 1999. An analysis of Monte Carlo integer programming, simulated annealing and tabu search for solving spatial harvest scheduling problems. *Forest Science* 45: 292-301.

Boston, K. and Bettinger, P. 2001. Development of spatially feasible forest plans: a comparison of two modelling approaches. *Silva Fennica* 35(4): 425-435.

Dykstra, D. 1984. *Mathematical programming for natural resource management*. McGraw-Hill, Inc. 318 p.

Falcão, A. and Borges, J. 2002. Combining random and systematic search heuristic procedures for solving spatially constrained forest management scheduling problems. *Forest Science* 48: 608-621.

Kangas, J. and Pukkala, T. 1992. A decision theoretic approach to goal programming problem formulation: an example on integrated forest management. *Silva Fennica* 26(3): 169-176

Kangas, J., Loikkanen, T., Pukkala, T. and Pykäläinen, J. 1996. A participatory approach to tactical forest planning. *Acta Forestalia Fennica* 251. 24 p.

Öhman, K. and Eriksson, L.O. 2002. Allowing for spatial consideration in long-term forest planning by linking linear programming with simulated annealing. *Forest Ecology and Management* 161: 221-230.

Pukkala, T. and Kangas, J. 1993. A heuristic optimisation method for forest planning and decision-making. *Scandinavin Journal of Forest Research* 8: 560-570.

Pukkala, T., Nuutinen, T. and Kangas, J. 1995. Integrating scenic and recreational amenities into numerical forest planning. *Landscape and Urban Planning* 32: 185-195.

Reeves, C. R. (ed.) 1993. *Modern heuristic techniques for combinatorial problems*. John Wiley & Sons, Inc. 320 p.

Render, B. and Stair, R.M. 1992. *Introduction to Management Science*. Allyn and Bacon. 856 p.

Saaty, R. 1980. *The Analytic Hierarchy Process. Planning, priority setting, resource allocation*. McGraw-Hill Publishing Company. 283 p.

Chapter 2

MEASUREMENT OF PREFERENCES IN MULTIPLE CRITERIA EVALUATION

JUHA M. ALHO
University of Joensuu, Department of Statistics, P.O. Box 111, FIN-80101 Joensuu, Finland

PEKKA KORHONEN
Helsinki School of Economics, Department of Economics and Management Science, Runeberginkatu 22-24, FIN-00100 Helsinki, Finland

PEKKA LESKINEN
Finnish Forest Research Institute, Kannus Research Station, P.O. Box 44, FIN-69101 Kannus, Finland

Abstract. In this paper, we deal with the problem of modelling preferences in multiple criteria evaluation situations. When the number of objects to be evaluated is small, then it is possible to make a detailed analysis of the decision-maker's preferences to find out a "value" or a "score" for each object. For example, in the Analytic Hierarchy Process, preference analysis is based on pairwise comparisons. We consider the statistical analysis of pairwise comparisons, and show that several issues of measurement scale must be clearly understood, before one can reliably apply the methods in practice. Our approach is based on the use of regression analysis rather than the eigenvalue technique of the AHP, to find the value scores for alternatives.

Keywords: preference modelling, Analytic Hierarchy Process, scaling, multi-attribute utility theory.

Timo Pukkala (ed.), Multi-objective forest planning, 21–36.

1. INTRODUCTION

In most decision and planning situations a decision maker will have to consider multiple criteria implicitly or explicitly. For a minor problem, the decision maker may not even realise the presence of multiple criteria in his/her evaluation. However, in major decision and planning situations it is important that the decision maker recognises all relevant criteria and uses them to evaluate the decision (or planning) alternatives.

The term Multiple Criteria Decision Making is used to refer to the solving of decision and planning problems that involve multiple, and often conflicting, criteria. "Solving" means that the decision maker chooses one "reasonable" alternative from among a set of available ones. In the cases we consider, the choice is assumed to be irrevocable. The chosen alternative is called the Most Preferred Solution. For such problems, it is typical that no unique solution exists. The solution is determined by the preferences of the decision maker instead.

Helping decision makers to deal with multiple criteria decision and planning problems has been subject to intensive studies since the beginning of the 1970s, but many theoretical concepts were defined much earlier. In the 1970s, the research focused on the theory of multiple objective mathematical programming. During the 1980s, a clear shift towards multiple criteria decision support occurred. Accordingly, much research has focused on the user interface, on the behavioural aspects of decision-making, and on supporting the entire decision-making process from problem structuring to solution implementation.

When the set of decision alternatives is finite and the elements are explicitly known in the beginning of the solution process, we refer to Multiple Criteria Evaluation Problems, Discrete Multiple Criteria Problems, or Selection Problems (for a survey, see, e.g., Olson 1996). When the number of alternatives is non-denumerable, the alternatives are usually defined using a continuous mathematical model, and we speak of Multiple Criteria Design Problems or Continuous Multiple Criteria Problems.

The multiple criteria nature is very typical for forestry planning problems, and for those relating to other natural resources and to landscape. Nevertheless, few papers dealing with those problems appeared before the early 1980s. An early example is the paper by Chang and Buongiorno (1981). They studied a resource allocation problem with the aim of developing a methodology for multiple use planning of publicly owned forests. They developed a mathematical model for this problem by combining goal programming and input-output analysis. Currently, most forest planning problems in which multiple criteria approaches are used belong to the class of multiple criteria evaluation problems.

For example, m items being compared might be forest plans suggested for a particular forest holding, and their relative merits need to be compared. The criteria with respect to which an assessment is desired may include future environmental impacts hard to quantify such as biodiversity, or the beauty of a landscape. Such criteria are often inherently multidimensional, so an evaluation of the net effect is called for. This may not be an easy task, especially since empirical data may be scanty and an element of forecasting the future state of the forest is typically involved.

In this paper, we will consider multiple criteria evaluation problems and discuss their applicability to forest research. Especially, we deal with a special class of those problems that have few alternatives, but possibly many hierarchical criteria. It is typical that not only the most preferred solution is sought, but the decision maker's preferences over all alternatives are analysed. In this paper, we use the framework of the AHP (the Analytic Hierarchy Process) developed by Saaty (1980) as a potential approach for the modelling of the preferences. In that technique, the information about the decision maker's preferences is based on pairwise comparison. The information need not be exact. Imprecise, and even inconsistent, evaluations are allowed. The solution procedure will "correct" inexactness.

Another example of a method based on detailed preference analysis is the Outranking Method by Roy (1973). Both methods are well known and widely used. The presence of multiple criteria is an essential feature in both, but the structuring of the problem is equally important.

In Section 2, we provide a brief introduction to multiple criteria evaluation problems. In Section 3, we review the regression-based approach to preference modelling, and in Section 4, discuss the effects of different scales. Discussion and concluding remarks are given in Section 5.

2. PRINCIPLES FOR SOLVING MULTIPLE CRITERIA EVALUATION PROBLEMS

A multiple criteria evaluation problem can be characterised as a vector maximisation problem in which a decision maker tries to maximise all components of vector $\mathbf{q} = (q_1, q_2, \ldots, q_p)$ simultaneously in a subset $Q \subset \Re^p$ consisting of a finite number of elements that are explicitly known in the beginning of the solution process. Therefore, we have a Multiple Criteria Evaluation Problem.

There are several alternative approaches that can help the decision maker to solve his/her problem. The choice depends on the characteristics of the problem. The outranking approach (Roy 1973), the multiattribute utility theory (MAUT) (Keeney and Raiffa 1976), the analytic hierarchy process (Saaty 1980), the regime method (Hinloopen et al. 1983), the interactive programming approach (Korhonen et al. 1984), the hierarchical interactive approach (Korhonen 1986), the visual reference direction approach (Korhonen 1988, Korhonen and Karaivanova 1999), and the aspiration-level interactive method (AIM) (Lotfi et al. 1992) are examples of the approaches developed to solve evaluation problems. A simple cross-tabulation made on the basis of the number of criteria and alternatives is given in Table 1.

Table 1. A classification of multiple criteria evaluation methods.

Number of alternatives	*Number of criteria*: *Small*	*Number of criteria*: *Large*
Small	• Outranking Method (Roy 1973) • MAUT (Multi-attribute utility theory) (Keeney and Raiffa 1976) • Regime method (Hinloopen et al. 1983)	• AHP (Analytic Hierarchy Process) (Saaty 1980) • HIRMU (Hierarchical Interactive Method) (Korhonen 1986)
Large	• Convex cone approach (Korhonen et al. 1984) • VIMDA (Visual Interactive Method for Discrete Alternatives) (Korhonen 1988) • AIM (An Aspiration-Level Interactive Model) (Lotfi et al. 1992)	

There are many other ways to classify the problems as well. For instance, in the outranking approach by Roy (1973), there is no need to explicitly specify the criteria at all. The preference relations between alternatives are the main object of interest. Those preferences can be analysed without explicit criteria as well. The multi-attribute utility theory by Keeney and Raiffa (1976) is able to deal with uncertainty in criterion values, and the analytic hierarchy process by Saaty (1980) is developed to handle the hierarchical structure of criteria.

More formally, our problem may be regarded as a *value* (or *utility*) *function* maximisation program

$$\max v(\mathbf{q}) \text{ s.t. } \mathbf{q} \in Q, \tag{1}$$

where v is strictly increasing in each argument and real-valued, and defined at least in the feasible region Q. Function v specifies the decision maker's preference structure. Without loss of generality, we assume that $v(\mathbf{q}) \geq 0$ for all $\mathbf{q} \in Q$. Generally, in multiple criteria decision problems, v is assumed to be unknown. Interactive approaches are applicable for those problems (see, e.g. Korhonen et al. 1984, Korhonen 1988, and Lotfi et al. 1992).

Typically, those problems consist of many alternatives, making the explicit comparison of alternatives impossible.

Solutions to our problem are alternatives that can be the solutions of (1) for some value function $v: Q \rightarrow \Re$. Those solutions are called *non-dominated.* Any choice from among the set of non-dominated (efficient) solutions is an acceptable and "reasonable" solution, unless we have additional information about the decision maker's preference structure.

Nondominated solutions are defined as follows:

Definition 1. An element $\mathbf{q}^* \in Q$ is *non-dominated* iff there does not exist another $\mathbf{q} \in Q$ such that $\mathbf{q} \geq \mathbf{q}^*$ and $\mathbf{q} \neq \mathbf{q}^*$ (i.e. $q_i \geq q_i^*$, $q_i \neq q_i^*$, $i = 1, 2, \ldots, p$).

Definition 2. An element $\mathbf{q}^* \in Q$ is *weakly nondominated* iff there does not exist another $\mathbf{q} \in Q$ such that $\mathbf{q} > \mathbf{q}^*$ (i.e. $q_i > q_i^*$, $i = 1, 2, \ldots, p$).

There are methods in which v is explicitly constructed by means of preference information received from a decision maker. In this case, a "value" or a "score" is attached to each alternative making possible, for instance, the ranking of alternatives. The multiple attribute utility theory (MAUT) is a classical example (Keeney and Raiffa 1976). Typically, in an MAUT approach, the strictly increasing marginal value (utility) functions $v_i(q_i) \geq 0$, $i = 1, 2, \ldots, p$, are first determined and a value function is introduced over all criteria by means of those marginal functions, $v(\mathbf{q}) = v(v_1(q_1), v_2(q_2), \ldots, v_p(q_p))$. An additional feature for MAUT is the presence of uncertainty, but that aspect is beyond the scope of this paper. Many approaches assume that the overall value function is a linear function of the marginal functions, i.e.,

$$v(\mathbf{q}) = \sum_{i=1}^{p} w_i v_i(q_i), \tag{2}$$

where $w_i > 0, i = 1, 2, \ldots, p$.

The above aggregation rule is used in AHP and in MAUT, for example. One more example is MACBETH (Bana e Costa and Vansnick 1999). Another typical way to aggregate the overall value function is to use a geometric mean, i.e.

$$v(\mathbf{q}) = \sqrt[n]{\prod_{i-1}^{p} v_i(\mathbf{q})^{w_i}}, \; n = \sum_{i=1}^{p} w_i. \tag{3}$$

Many authors (see, e.g., Barzilai et al. 1987, and Lootsma 1991) have proposed a multiplicative version as a variant to the standard AHP.

Because Q consists of a finite number of elements, a basic data set of a multiple criteria evaluation problem can be also given in an $m \times p$ matrix $A \in \Re^{m \times p}$ whose elements a_{ij}, $i \in I = \{1, 2, ..., m\}$ and $j \in J = \{1, 2, ..., p\}$ represent the criterion values on alternatives. We often use simply index i to refer to the alternative in row i. Occasionally, we may use capital letters A, B, C, ... to refer to alternatives as well. Actually, the specification of marginal value functions attaches a value score to each alternative on each criterion. Thus we could specify the problem by means of those scores v_{ij}, $i = 1, 2,\ldots, m$ and $j = 1, 2,\ldots, p$.

In the approaches based on Equations 2 and 3, the user has to answer the following questions:

1 How will marginal values be determined?
2 How will those values be scaled?
3 How will the weights be determined?
4 How will the weights be scaled?

To find the answers to those questions we have to model a decision maker's preferences. To do this, various behavioural assumptions are needed. A common assumption is that when a decision maker compares alternatives and expresses the result of an evaluation, he/she uses a ratio scale. If the decision maker says for instance that "Alternative A is much better than alternative B" and "C is better than D", then we interpret that to mean: $v(A)/v(B) > v(C)/v(D)$, $v(B) > 0$, $v(D) > 0$.

Another important assumption concerns the interpretation of weights w_i of the criteria, $i = 1, 2,\ldots, p$. In the AHP, MAUT, etc. those weights are often interpreted to describe the relative importance of the criteria. If, e.g., we assume that $w_k v_k(q_k(i)) = w_h v_h(q_h(m)) > 0 \Rightarrow w_k/w_h = v_h(q_h(m))/v_k(q_k(i))$, where indices k and h refer to criteria and $q_k(i)$ and $q_h(m)$ refer to the values of those criteria on alternatives i and m, respectively. However, the articulated importance does not necessarily mean that the value ratio is equal to the inverse of the weight ratio. The decision maker may, for instance, consider some criterion very important and desire to achieve a certain aspiration value on that specific criterion. In addition, the way the marginal value functions are scaled has a strong impact on results. For instance in MAUT, $\max v_k(q_k(i)) = 1$, $i = 1, 2, ..., m$, $\forall k$. The standard AHP-technique uses the scaling $\Sigma_i v_k(q_k(i)) = 1$.

This is an ongoing debate between the AHP and MAUT schools on the relative superiority of those scaling methods. The supporters of a multiplicative model bring their own flavours to this debate.

In this paper, we will not deal with those aspects. We will consider how to determine value scores for all alternatives on a certain criterion when information from the decision maker is gathered via pairwise comparison.

3. STATISTICAL ANALYSIS OF RATIO SCALE ASSESSMENTS

In this section, we will not use the term "decision maker" to refer only to a decision maker, but also to an expert involved in the decision process. The decision maker may be a forest owner who wants to systematise his/her beliefs concerning decision alternatives. Or, the decision maker may be an expert, whose views are being solicited to support actual decision-making. There may also be several decision makers, as in the case of estimating the aesthetic preferences of nature park user.

A possible approach to decision making is to simply order the items according to the decision maker's preferences. This has the virtue of simplicity and consistency. The rank order can be further refined by assigning scores to the items, in order to quantify the differences between consecutive items. Unfortunately, the process may be sensitive to the order the items are presented. In addition, the relative merits of two items may crucially depend on what other items are being included in the set of items.

Systematic pairwise comparisons have been suggested as a means of deriving potentially more robust preferences from decision makers. A central element of the so-called analytic hierarchy process (AHP) is a full set of $m(m-1)/2$ pairwise comparisons (Saaty 1977, 1980). When the items 1 and 2 are compared in a pairwise manner, one hopes that the effect of other competitors $(i = 3,\ldots, m)$ has less effect on the result than in a direct, overall evaluation of all items at once. Moreover, the quantification of the relative merits of 1 and 2 is not, at least directly, dependent on the scores the other items may have been assigned. The cost of these benefits is that the comparison process may be cumbersome. Moreover, there is no guarantee that the resulting pairwise assessments are consistent with each other. For example, if one prefers item 1 to item 2 by 2:1 and item 2 to item 3 by 3:1, one should prefer item 1 to item 3 by 6:1 to be consistent.

Fortunately, it is possible to use statistical regression analysis to alleviate the problems. A set of full pairwise comparisons is not necessary for estimation; thus, the common problem of missing data in multiple decision maker settings needs not invalidate the study. Moreover, one can make a virtue out of the (likely) inconsistency, because regression statistically assesses the uncertainty of the elicited views. One can then take the uncertainty into account when the decision maker's views are used in decision support.

A potential weak point of any quantitative comparison of multidimensional decision alternatives is the assignment of scores. Since a decision problem can always be embedded in a larger problem with potentially more advantageous items, absolute scores cannot be determined. A more modest goal is to assign relative scores such that the m items being compared are correctly represented with respect to each other. As noted by Saaty (1977), it is difficult for a person to distinguish simultaneously more than 7–9 different levels of preference. Saaty used the following verbal descriptions and corresponding scores:

Score	*Description*
1	equal importance
3	weak importance of one over the other
5	essential or strong importance
7	demonstrated importance
9	absolute importance

Intermediate scores of 2, 4, 6, and 8 can also be assigned. For example, if a person thinks that item 1 is at least weakly better than item 2, but (s)he is not comfortable with saying that 1 is strongly better than 2, then his or her view might be represented by the score 4. If item 2 is at least weakly but not necessarily strongly better than item 1, then the score 1/4 would be assigned to the comparison of 1 to 2. Alternative systems of scoring will be discussed below.

More formally, let the ordered pair (i, j) denote the comparison of i to j. Let the relative score of i to j be $r_{ij} > 0$. In the classical AHP procedure one forms the $m \times m$ matrix $\mathbf{R} = (r_{ij})$ by first eliciting the values r_{ij} for $i < j$, and then defining $r_{ii} = 1$ and $r_{ji} = 1/r_{ij}$ for $i > j$. In this manner a matrix of positive elements is obtained. By the Perron-Frobenius theorem (Gantmacher 1959) $\mathbf{R}$ has a positive eigenvalue λ that is strictly greater than the absolute value of the other eigenvalues. The elicited views are *consistent*, if $r_{ij}r_{jk} = r_{ik}$ for all i, j, k. To the dominant eigenvalue there corresponds an eigenvector $\mathbf{w} = (w_1,\ldots, w_m)^{\mathrm{T}}$, or $\mathbf{Rw} = \lambda\mathbf{w}$ that can be chosen to have $w_i > 0$ and $w_1 +\ldots+ w_m = 1$. The value w_i can then be interpreted as the relative value of item i, because for a consistent set of elicited views we have that $\Sigma_j\, r_{ij}w_j = mw_i$. In fact, one can prove that in general $\lambda \geq m$ with equality only if the comparisons are exactly consistent (Saaty 1977). As discussed by Saaty and Vargas (1984), other estimation criteria, such as least squares or logarithmic least squares, could also be used. The regression formulation to be discussed below relies on the latter.

A practical drawback of the eigenvector formulation is that it requires all $m(m-1)/2$ comparisons to be made with no missing values. Another problem is that if there are other eigenvalues that are large in absolute value, then the approximation $\mathbf{R} \approx \lambda\mathbf{wu}^{\mathrm{T}}$, where $\mathbf{u} = (1/w_1,\ldots, 1/w_m)^{\mathrm{T}}$, can be poor. This can happen if the elicited views are inconsistent with each other. *Ad hoc* rules of thumb have been developed to assess the consistency of the elicited views (e.g., Saaty 1977) based on the difference $\lambda - m$. However, there is no systematic statistical theory that could be used as a foundation for the measures. The regression approach provides an answer to all these problems.

Suppose the value of item i is $v_i = \exp(\mu+\alpha_i)$, $i = 1,\ldots, m$. Then the relative value of i to j is $\exp(\alpha_i-\alpha_j)$. Suppose the actual elicited relative preference r_{ij} is of the form $r_{ij} = \exp(\alpha_i-\alpha_j+\varepsilon_{ij})$, where the random errors have the expectations $\mathrm{E}(\varepsilon_{ij}) = 0$. Note that in this formulation we think of error in relative terms. Defining $y_{ij} = \log(r_{ij})$, we get the regression model

$$y_{ij} = \alpha_i - \alpha_j + \varepsilon_{ij} \tag{4}$$

for $i < j = 2,\ldots, m$. For identifiability we set $\alpha_m = 0$, for example. The regression formulation has been investigated by De Jong (1984), Crawford and Williams (1985), Carriere and Finster (1992), and Zhang and Genest (1996), for example. Our discussion follows primarily those in Alho et al. (1996), and Alho et al. (2001), where many additional details and extensions can be found.

By coding the $m(m-1)/2$ observations y_{ij} into a vector $\mathbf{Y}$, the error terms ε_{ij} in the same order into a vector ε and by defining the corresponding design matrix $\mathbf{X}$ (for an example, see, e.g., Alho et al. 1996), we can write the regression model in the matrix form $\mathbf{Y} = \mathbf{X\alpha}+\boldsymbol{\varepsilon}$, where $E(\varepsilon) = \mathbf{0}$. If the errors are uncorrelated and homoscedastic with $Cov(\varepsilon) = \sigma^2\mathbf{I}$, the ordinary least squares solution $\hat{\boldsymbol{\alpha}} = (\mathbf{X}^T\mathbf{X})^{-1}\mathbf{X}^T\mathbf{Y}$ is optimal. More generally, a generalised least squares solution is available. The estimates $\hat{\alpha}_i$ can be transformed to a commonly used priority scale by $\hat{a}_i = \exp(\hat{\alpha}_i)/\Sigma_i \exp(\hat{\alpha}_i)$ so that $\Sigma_i \hat{a}_i = 1$. It is immediately clear from the regression equation that if each item is involved in at least one pairwise comparison, then a solution exists.

Given the difficulty of expressing precise views of such matters as future biodiversity or aesthetic beauty, one expects that the elicited views are, to some extent, inconsistent. In the regression formulation the residual variance $\hat{\sigma}^2 = \|\mathbf{Y} - \mathbf{X}\hat{\boldsymbol{\alpha}}\|^2/(n-m+1)$, where n is the number of pairwise comparisons, is a natural estimate of inconsistency. In particular, it can be directly translated into an estimate of uncertainty of the estimates via $\mathrm{Cov}(\hat{\boldsymbol{\alpha}}) = \hat{\sigma}^2(\mathbf{X}^T\mathbf{X})^{-1}$. Under a normal assumption, standard inferential tools such as t and F tests become available. Bootstrap is also available, if analytical assumptions seem unwarranted. In Alho and Kangas (1997) a Bayesian formulation was also provided under a diffuse prior. In some applications, the Bayesian interpretation of the results is more natural than the classical frequentist interpretation. Software written as MATHEMATICA functions is also available on the web site of the University of Joensuu (http://joyx.joensuu.fi/~koleh/osmo2.html).

The regression formulation can be extended in many ways. As discussed in Alho et al. (2001), we can add explanatory variables describing either the decision makers or the items, to explain the elicited views. The advantage of these approaches over first running an eigenvalue solution and then using the resulting scores as variables to be explained in regression, is that in the regression formulation both analyses are carried out under the same set of assumptions. Similarly, techniques of experimental design can be used to plan pairwise comparisons experiments.

4. COMPARISON OF RATIO SCALE MEASUREMENT MODELS

In the AHP literature, several measurement scales have been proposed for the numerical elicitation of the pairwise comparisons r_{ij} in ratio scale (Fig. 1). In addition to the original measurement scale of Saaty (1977), we introduce in this section the measurement scales of Lootsma (1993), Salo and Hämäläinen (1997), and Ma and Zheng (1991). Then we write the measurement scales as a function of a scale parameter describing the magnitude of the numerical ratios (Leskinen 2001). This enables us to separate the principle of constructing a measurement scale from its numerical values. Finally, we review the results concerning the statistical properties of the measurement scales (Leskinen 2001).

In the original AHP method, a discrete and verbal measurement scale is used with numerical counterparts 1/9, 1/8,..., 1/2, 1/1, 2/1, ..., 8/1, 9/1. For example, r_{ij} = 1/9 means that the value of attribute i is equal to 1/9 times the value of attribute j. Consider the r_{ij}, and the corresponding verbal judgments denoted by δ_{ij} = 0,..., 8, where 0 means "attributes i and j are equal", ..., and 8 means "attribute i is absolutely preferred over attribute j". If attribute j is preferred to attribute i, then $1/r_{ij}$ can be used instead of r_{ij}. Because Saaty used an arithmetic sequence of numbers, the original measurement scale is turned to model (Leskinen 2001)

$$r_{ij} = s\delta_{ij} + 1, \tag{5}$$

where $s > 0$ is the scale parameter. For example, s=1 leads to the original values 1/1, 2/1, 3/1, ..., 9/1. If s=2, then we have the ratios 1/1, 3/1, 5/1, ..., 17/1 etc.

Lootsma (1993) argued that the numerical counterparts should follow a geometric progression, and proposed a measurement scale of the form $\exp(s\eta_{ij})$, where $s > 0$ is the scale parameter and η_{ij} = –8,..., 0,..., 8 is an index for the verbal judgments. For example, –8 means that "attribute j is absolutely preferred to attribute i". To calculate the priorities of decision alternatives with respect to decision criteria, for example, Lootsma suggested s = 0.693, which gives the sequence 1/256, 1/128,..., 1/2, 1/1, 2/1,..., 128/1, 256/1. By using the notation in Equation 5, we can also write the geometric model as (Leskinen 2001)

$$r_{ij} = \exp(s\delta_{ij}). \tag{6}$$

Salo and Hämäläinen (1997) considered that the measurement scale should lead to an even dispersion on the priority scale, and proposed the numerical values 1/9, 1/5.67, ..., 1/1.22, 1/1, 1.22/1, ..., 5.67/1, 9/1. This means that (Leskinen 2001)

$$r_{ij} = \frac{s\delta_{ij} + 0.5}{1 - (s\delta_{ij} + 0.5)}, \tag{7}$$

where $0 < s < 0.0625$. For example, $s = 0.05$ gives the values proposed by Salo and Hämäläinen. The model leads to an even dispersion on the scale of priorities in the sense that $a_i = s\delta_{ij} + 0.5$ in a single pairwise comparison. Because $a_i < 1$, we need the upper limit $s < 0.0625$. Finally, Ma and Zheng (1991) pointed out that the numerical values should agree with the verbal expressions, and proposed the "9/9 - 9/1" - principle, i.e. values 1/9, 2/9, ..., 8/9, 9/9, 9/8, ..., 9/2, 9/1. This implies the model (Leskinen 2001)

$$r_{ij} = \frac{s}{s - \delta_{ij}(s-1)/8}, \tag{8}$$

where $s > 1$. For example, the original numerical values can be obtained by $s = 9$.

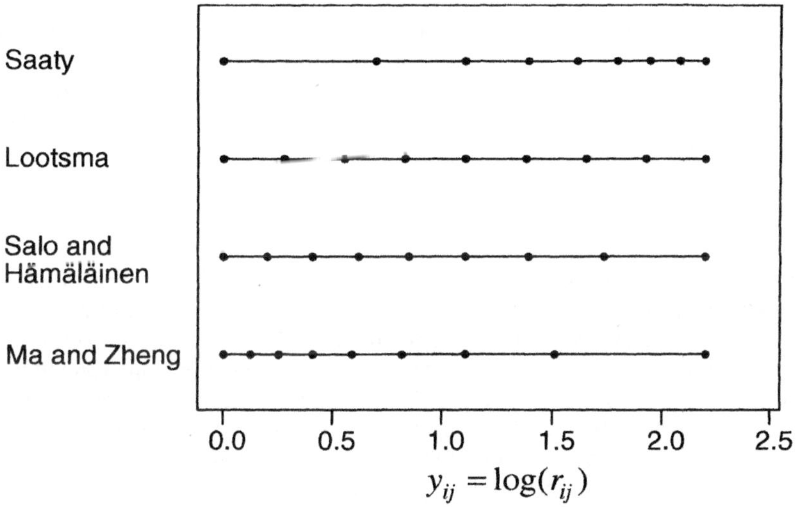

Figure 1. Numerical values of four measurement scales from $r_{ij} = 1/1$ to $r_{ij} = 9/1$.

Leskinen (2001) compared the models in Equations 5–8 by the regression analysis of pairwise comparisons data. In the arithmetic measurement scale of Equation 5, for example, we will have a regression model

$$\log(s\delta_{ij} + 1) = \alpha_i - \alpha_j + \varepsilon_{ij}. \tag{9}$$

Consider first a measurement scale on which the number of different numerical values is the same as the number of unequal attributes being compared. Leskinen (2000, 2001) showed that in this case only the geometric measurement scale of Equation 6 enables consistent r_{ij}. Consequently, all other measurement scales of Equations 5–8 will produce inconsistent data regardless the judgements. The result holds for any $s > 0$ in the geometric measurement scale. On the other hand, if the number of different numerical values is larger than the number of unequal attributes being compared, it is possible to have consistency also by other measurement scales. For example, Equation 5 with scale parameter $s = 1$ allows $r_{12} = 2/1$, $r_{23} = 3/1$, and $r_{13} = 6/1$. However, Leskinen (2000, 2001) showed that the possibilities of producing consistent data are small when using Equations 5, 7 or 8 with the scale parameters that lead to the original measurement scales proposed by Saaty, Salo and Hämäläinen, and Ma and Zheng. The conclusion was that the observed inconsistency (i.e. $\hat{\sigma}^2 > 0$ in the regression model) can be interpreted on the geometric measurement scale as the uncertainty of the decision maker, because consistent judgments would have been possible (Leskinen 2000, 2001). On other measurement scales, the inconsistency can be also caused by the measurement scale itself.

One problem of the ratio scale assessments is that the numerical interpretations of pairwise comparisons is context- and decision-maker-specific. Leskinen (2000, 2001) studied this problem through the model

$$s\delta_{ij} = \alpha_i - \alpha_j + \varepsilon_{ij} \tag{10}$$

by considering the impact of the scale parameter on the statistical inference, when the geometric measurement scale is used. It turned out that the t test for the null hypothesis H_0: $\alpha_i = \alpha_j$ based on Equation 10 is independent of s (Leskinen 2000, 2001). This provides a robust way to make decisions independently of the numerical interpretation of verbal pairwise comparisons. The result can be extended to decision hierarchy under some conditions (Leskinen 2000, 2001). For example, the scale independence presumes the use of a geometric-mean aggregation rule instead of the arithmetic-mean aggregation rule used in the original AHP (Saaty 1977). However, the numerical values of the estimated priorities depend on s in each case. In general, large scale parameters will lead to large differences of priorities, and vice versa.

Leskinen (2001) compared the differences between Equation 5–8 in the context of statistical inference, in addition to that of inconsistency. For comparability, the scale parameters that lead to 9/1 as the largest numerical value were used with all four models. Consider the hypothesis H_0: $\alpha_1 = \alpha_2$ against H_1: $\alpha_1 \neq \alpha_2$ based on models like Equations 9 and 10. The simulation results concerning the population of different sets of pairwise comparisons showed that the acceptance or the rejection of H_0 is sensitive to the choice of the measurement scale (Leskinen 2001). Moreover, the measurement scales of Saaty, and of Ma and Zheng often produce either the largest or the smallest p value among the four measurement scales. This means that the geometric measurement scale, and that of Salo and Hämäläinen, are more robust in that the hypothesis tests are not as affected by the misspecification of the measurement scale.

The models in Equations 5–8 were compared also by organising a pairwise comparisons experiment with known true values v_i/v_j by asking the decision makers for pairwise comparisons of, e.g., length of figures (Leskinen 2001). This enables us to study the accuracy of the measurement scales according to how close the estimates $\exp(\hat{\alpha}_i)/\exp(\hat{\alpha}_j)$, based on alternative measurement scales, are to the v_i/v_j. The comparison was done by first estimating the optimal scale parameters for each measurement scale so that the performance of Equations 5–8 can be studied under optimal scaling (Leskinen 2001). An alternative approach would be to use fixed scale parameters, but in that case non-optimal scaling can dominate the results. For the arithmetic measurement scale of Equation 5, for example, the scale parameter was estimated from the model (Leskinen 2001)

$$\log(v_i / v_j) = \log(s\delta_{ij} + 1) + e_{ij}, \tag{11}$$

where e_{ij} is the residual term. Also combinations of the original measurement scales were tested in order to study the performance of the measurement scales in the area between the original models. For example, the combination of Equations 5 and 6 leads to (Leskinen 2001)

$$\log(v_i / v_j) = \log(s_1\delta_{ij} + 1) + s_2\delta_{ij} + e_{ij}, \tag{12}$$

where $s_1 > 0$, and $s_2 > 0$.

According to the results concerning pooled data of several decision makers, the least squares estimate of the scale parameter in geometric measurement scale produced the smallest estimate for $\mathrm{Var}(e_{ij})$, but the differences compared to the other measurement scales were small (Leskinen 2001). Moreover, the combination models of type Equation 12 provided only small improvement compared to the individual models of type Equation 11. This means that there were no better models available in the area between the original measurement scales. The optimality of the geometric model suggests also that there are no better models available outside the range of the original models for multiple decision maker data, because the geometric model is not an extreme one in the set of the four models considered (Leskinen 2001). On the other hand, more extreme models can be useful for single decision maker data, but the effects of misspecification of the model would be large. Finally, the accuracy of the estimates $\exp(\hat{\alpha}_i)/\exp(\hat{\alpha}_j)$ obtained from models like Equations 9 and 10 depends on the optimality of the scale parameter, but is relatively independent of the functional form of a measurement scale (Leskinen 2001).

In summary, the above results support the use of the geometric model of Equation 6 in the ratio scale assessment of preferences by pairwise comparisons technique (Leskinen 2001). The advantages of the geometric model include reasonable interpretation of observed inconsistency (see also Leskinen and Kangas 1998), and robustness against the misspecification of the measurement scale. Moreover, the geometric measurement scale has the additional property of scale independent statistical inference under some conditions. On the other hand, the accuracy of the estimated preferences is relatively independent of the mathematical form of the measurement scale, but depends highly on the optimality of the scale parameter. This, and the fact that also statistical inference can depend on the scale parameter, will leave us the problem of determining the value of s in Equation 6 for scale dependent decision support problems. One solution would be to assess the numerical range of the geometric model before the actual pairwise comparisons (Leskinen 2001).

5. CONCLUDING REMARKS

In this paper, we have considered the use of pairwise comparisons as an evaluation technique with the aim of finding value scores for alternatives. Pairwise evaluation was based on the assumption that the decision maker uses a ratio scale in his/her evaluation. Various potential measurement scales were also studied. The so-called geometric model worked best in our experiments, but the problem of determining the value of the scale parameter of the geometric model still remains for scale dependent decision support problems.

Our discussion shows that there are several aspects in the coding of the verbal answers that can influence the conclusions. For the successful application of the methods based on pairwise comparison techniques, it is essential that practitioners become aware of the problems, and recognize the added uncertainty in one way or another. A useful approach is to try out alternative scales and see if this makes a difference.

The ultimate goal is to use the scale and the method, which helps to find the "real" value scores for alternatives. However, it is extremely difficult if not impossible to evaluate the reaching of this goal. The value scale we are able to find always depends on the method we use. The method as such also affects the decision maker's preferences. In this paper, we have studied the problem from a different perspective. We have assumed that the scale is known and asked which kind of pairwise comparison technique makes it possible to find an original scale.

REFERENCES

Alho, J.M., Kangas, J. and Kolehmainen, O., 1996: Uncertainty in expert predictions of the ecological consequences of forest plans. *Applied Statistics* 45: 1-14.

Alho, J.M. and Kangas, J., 1997: Analyzing uncertainties in experts' opinions of forest plan performance. *Forest Science* 43: 521-528.

Alho, J.M., Kolehmainen, O. and Leskinen, P., 2001: Regression methods for pairwise comparisons data. In D.L. Schmoldt, J. Kangas, G.A. Mendoza and M. Pesonen (eds.), *The Analytic Hierarchy Process in Natural Resource and Environmental Decision Making*. Kluwer Academic Publishers, p. 235-251.

Bana e Costa, C.A. and Vansnick, J.C., 1999: The MACBETH approach: basic ideas, software and an application. In N. Meskens and M. Roubens (eds.), *Advances in Decision Analysis*. Kluwer Academic Publishers, p. 131-157.

Barzilai, J., Cook, W. and Golany, B., 1987: Consistent weights for judgements matrices of the relative importance for alternatives. *Operations Research Letters* 6: 131-134.

Carriere, J. and Finster, M., 1992: Statistical theory for the ratio model of paired comparisons. *Journal of Mathematical Psychology* 36: 450-460.

Chang, S. and Buongiorno, J., 1981: A programming model for multiple use forestry. *Journal of Environmental Management* 13: 45-58.

Crawford, G. and Williams, C., 1985: A note on the analysis of subjective judgment matrices. *Journal of Mathematical Psychology* 29: 387-405.

De Jong, P., 1984: A statistical approach to Saaty's scaling method for priorities. *Journal of Mathematical Psychology* 28: 467-478.

Gantmacher, F.R., 1959: *The Theory of Matrices*. Vol. II, Chelsea, New York.

Hinloopen, E., Nijkamp, P. and Rietveld, P., 1983: The regime method: a new multicriteria technique. In P. Hansen (ed.), *Essays and Surveys on Multiple Criteria Decision Making*. Springer, p. 146-155.

Keeney, R. and Raiffa, H., 1976: *Decisions with Multiple Objectives*. Wiley, New York.

Korhonen, P., Wallenius, J. and Zionts, S., 1984: Solving the discrete multiple criteria problem using convex cones. *Management Science* 30: 1336-1345.

Korhonen, P., 1986: A hierarchical interactive method for ranking alternatives with multiple qualitative criteria. *European Journal of Operational Research* 24: 265-276.

Korhonen, P., 1988: A visual reference direction approach to solving discrete multiple criteria problems. *European Journal of Operational Research* 34: 152-159.

Korhonen, P. and Karaivanova, J., 1999: An algorithm for projecting a reference direction onto the nondominated set of given points. In *IEEE Transactions on Systems, Man, and Cybernetics - Part A: Systems and Humans* 29: 429-435.

Leskinen, P. and Kangas, J., 1998: Analysing uncertainties of interval judgment data in multiple-criteria evaluation of forest plans. *Silva Fennica* 32: 363-372.

Leskinen, P., 2000: Measurement scales and scale independence in the Analytic Hierarchy Process. *Journal of Multi-Criteria Decision Analysis* 9: 163-174.

Leskinen, P., 2001: *Statistical Methods for Measuring Preferences*. University of Joensuu, Publications of Social Sciences 48.

Lootsma, F.A., 1991: *Scale Sensitivity and Rank Preservation in a Multiplicative Variant of the AHP and SMART*. Report 91-67, Faculty of Technical Mathematics and Informatics, Delft University of Technology, Delft, The Netherlands.

Lootsma, F.A., 1993: Scale sensitivity in the multiplicative AHP and SMART. *Journal of Multi-Criteria Decision Analysis* 2: 87-110.

Lotfi, V., Stewart, T.J. and Zionts, S., 1992: An aspiration-level interactive model for multiple criteria decision making. *Computers and Operations Research* 19: 671-681.

Ma, D. and Zheng, X., 1991: 9/9 - 9/1 scale method of AHP. *In Proceedings of the 2nd International Symposium on the AHP*. Vol. I, University of Pittsburgh, Pittsburgh, PA, p. 197-202.

Olson, D., 1996: *Decision Aids for Selection Problems*. Springer Series in Operations Research, New York.

Roy, B., 1973: How outranking relation helps multiple criteria decision making. In J. Cochrane and M. Zeleny (eds.), *Multiple Criteria Decision Making*. University of South Carolina Press, Columbia, SC, p.179-201.

Saaty, T.L., 1977: A scaling method for priorities in hierarchical structures. *Journal of Mathematical Psychology* 15: 234-281.

Saaty, T.L., 1980: *The Analytic Hierarchy Process. Planning, Priority Setting, Resource Allocation*. McGraw-Hill, New York.

Saaty, T.L. and Vargas, F., 1984: Comparison of eigenvalue, logarithmic least squares and least squares methods in estimating ratios. *Mathematical Modelling* 5: 309-324.

Salo, A.A. and Hämäläinen, R.P., 1997: On the measurement of preferences in the Analytic Hierarchy Process. *Journal of Multi-Criteria Decision Analysis* 6: 309-319.

Zhang, S.-S. and Genest, C., 1996: Etude d'un test de confirmation des priorités dans le cadre dur procédé d'analyse hiérarchique. *Revue de Statistique Appliquée* 44: 81-103.

Chapter 3

MULTIPLE CRITERIA DECISION SUPPORT METHODS IN FOREST MANAGEMENT

An overview and comparative analyses

JYRKI KANGAS
Finnish Forest Research Institute, Joensuu Research Station, P. O. Box 68, FIN-80101 Joensuu, Finland

ANNIKA KANGAS
University of Helsinki, Department of Forest Resource Management, P. O. Box 27, FIN-00014 Helsingin yliopisto, Finland

Abstract. Nowadays, forests are often managed for multiple uses. Forests are expected to produce reasonable incomes while at the same time promoting nature conservation and amenity values. There are also other characteristics that make natural resources decision-making situations complex. For example, group decision making and public participation are often required. To help decision makers make good choices, information and analyses are needed on the decision situation, on alternative choices of action, and on the consequences of alternative choices as well as on the preferences among these consequences. Multiple Criteria Decision Support (MCDS) methods are decision analysis tools that have been developed for dealing with all that information in order to support complex decision making with multiple objectives. In this chapter, some MCDS methods that recently have been applied to forestry or other natural resources management planning problems, and have been found to be promising, will be presented. In addition, some forestry applications are briefly described, and experiences gained using MCDS methods in forest management are discussed. Of the MCDS approaches, a closer look is taken at the Analytic Hierarchy Process, outranking methods, voting approaches, and the Stochastic Multicriteria Acceptability Analysis, because of their potentials for application. Applications for practically all MCDS methods with different qualities can be found in the field of natural resources decision support. However, no single method is best for all the decision support processes. The tool to be used should be chosen to fit the situation at hand; i.e., planning-case-wise consideration is always needed in order to build up an appropriate decision support process.

Keywords: Analytic Hierarchy Process, ELECTRE, forest planning, MCDS methods, Multicriteria Approval, forest management, outranking, PROMETHEE, SMAA, voting approaches

Timo Pukkala (ed.), Multi-objective forest management, 37–70.

1. INTRODUCTION

1.1. Need for multiple criteria decision support in forest management

The objective of forest management planning is to provide support for decision making so that an efficient mix of inputs and outputs best fulfilling the objectives set for the management of the area under planning can be found. Briefly put, the main phases in a strategic forestry decision-making process are: (1) data acquisition and assessing the present state of the area and its resources, (2) clarifying who are the decision makers and (in participatory approaches) other stakeholders involved and what are their objectives and preferences regarding the use of the area, (3) generating alternative management plans for the planning area and predicting their consequences, (4) evaluating and comparing the alternatives, and (5) choosing the best alternative among those deemed efficient, with respect to the objectives and preferences as clarified in phase (2). The phases do not necessarily proceed just in this order, and they might be applied iteratively, interactively and simultaneously.

The general framework of natural resources management, especially in forestry, has dramatically changed over the past ten years. Nowadays, forests are often managed for multiple uses. Forests should produce reasonable incomes while at the same time promoting, for instance, conservation and recreational considerations. This applies both to private non-industrial landowners as well as corporate, e.g., forestry-industry, and also to state and the non-corporate public sector. Generally, economic, ecological and socio-cultural sustainability are all required in the management of forests. Ecological and social sustainability have recently received much attention. Consequently, multi-functional forestry calls for multi-objective forest management planning and decision support.

In addition, there are increasing needs to take into account concerns and objectives of multiple decision makers or participants in decision making. This, in turn, calls for group decision support and participatory planning. In a case of multiple decision makers or participants, each might have different aims and objectives concerning the use of forest or other natural resources. The aims are often contradictory and they should somehow be made commensurable to find the best consensus or at least a good compromise solution.

There are also other characteristics that make forestry decision making situations complex. Compared to some other fields of decision making, they usually involve a high level of uncertainty about the variables. The consequences of the possible actions are often predicted using complex simulation models and/or expert opinions. In addition, the long time horizons further complicate the prediction of the consequences of implementing different management schedules. Long time horizons are needed especially when assessing the sustainability of alternative decisions, and actions have long-lasting effects on economic, ecological and socio-cultural considerations. Furthermore, there may also be a number of criteria measured in verbal or ordinal scale.

Research on forest planning has answered these challenges by applying and developing decision support methods and techniques for multiple criteria and participatory planning. Planning problems have been addressed by quantitative and qualitative approaches. Both approaches have their benefits and disadvantages (e.g., Kangas et al. 2001a). In practical planning processes, all the information available should be dealt with – regardless of whether the information is descriptive or numerical; subjective or objective; or based on preferences, expert judgements, measurements or forecasts. Consequently, participatory and multiple use forest planning call for analytical decision support methods and efficient decision analyses by which the various kinds of information can be dealt with in a versatile manner.

Strategic planning of forestry and other natural resources management is an excellent application area for methods of multiple criteria decision support (MCDS). The quality of forest planning can be improved by making use of the decision support methods, techniques, and tools produced by recent research, each having different characteristics and qualities. The key task is to choose appropriate approaches and methods for each decision support process. Often, the requirements and needs of forest decision support call for specifically tailored applications.

In this chapter, some MCDS methods that recently have been applied to strategic forest management planning and that have been found worth using or worth developing further will be presented. In addition, some examples of their forestry applications are briefly described, and experiences gained using MCDS methods in forest management are discussed. Of the MCDS approaches, a closer look is taken at the Analytic Hierarchy Process, outranking methods, voting approaches, and the Stochastic Multicriteria Acceptability Analysis; each having different methodological characteristics and thus being suitable for supporting different kinds of choice problems.

1.2. Fundamentals of decision analyses

The aim of decision analysis is to lend support to decision making involving problems that are too complex to be solved by intuition or common sense alone. Strategic forest management decisions are examples of such problems. In a decision-theoretic study, a decision is considered as a choice between two or more alternatives. In a normative approach to decision-making, the starting point is that a rational decision maker aims to choose the alternative that most probably maximises the decision maker's utility, determined on the basis of information available to him or her on the decision alternatives (Kangas 1992). This is the viewpoint in the case of a single decision maker. In group decision making, the total utility to be maximised can be taken as a combination of utilities of the persons belonging to the group. In participatory decision-making processes, some or even all the decision-making power might be allocated to the participants.

In decision support, information is produced on the decision situation, on alternative choices of action, and on the consequences of alternative choices, in order to help the decision makers make better decisions, the possible consequences of which they are aware. A complete decision model constitutes the basis for the decision support. Three things are represented in the decision basis: the alternatives available, information about the consequences associated with these alternatives, and the preferences among these consequences (Bradshaw and Boose 1990). Each aspect of the information has to be sound so that the best, a good, or at least a satisfactory alternative can be selected. Errors and defects in any part of the decision analysis can lead to questionable or invalid results.

In decision analysis, the decision situation is observed holistically. Generally, numerical encoding of information concerning the decision situation can be taken as a precondition for an effective and thorough analysis of a complex decision problem (von Winterfeldt 1988, Guariso and Werthner 1989). A numerical decision analysis is based on logical axioms and a methodology founded on these axioms and systematic procedures. In a numerical decision analysis, decision makers' and other stakeholders' preferences must be accommodated by the model.

A utility model, which can also be called a priority model, is a tool for evaluating the alternatives with respect to goals, objectives, opinions, etc. The utility model derived in the decision analysis is the key in combining the three parts of the decision basis. The utility – explicitly modelled or not – can be seen as an underlying basis of any rational choice. Often, the criteria for decision making are variables of the utility function and the parameters indicate the importance of the criteria. The aim is neither to explain observed behaviour nor to predict how decisions will be made, but to facilitate the making of better decisions than would be possible without the support. Although human behaviour might not be explained by models of rational behaviour, the preferences of decision makers can be analysed and the decision alternatives evaluated on the basis of these preferences, using an analytical decision model (Kangas 1993).

Utility is influenced by all attributes that have value to the decision maker. It is a measure of subjective desirability. Utility of a single decision maker can also include elements related to other people's preferences. So, maximising one's expressed utility is not necessarily purely selfish behaviour. In most cases, utility cannot be expressed in physical quantities; the real utility of physical units is determined by their value to the decision maker, and it is not always in a linear relation to the quantity of the units. In decision analyses, it is often better to use relative values instead of physical quantities (Forman 1987). However, in many practical decision processes the information available and its characteristics put limits on the quality of analyses.

If the utility model can arrange the decision alternatives only from the best to the worst, it determines ordinal utility. If the utility model can be interpreted on an interval or a ratio scale, it determines cardinal utility. In principle, it is sufficient to determine only ordinal utility when the best decision alternative is sought. Estimating the cardinal utility, however, enables versatile analyses of a complex decision situation. Cardinality on the ratio scale also makes thorough sensitivity and risk analyses possible (e.g., von Winterfeldt and Edwards 1986).

In most technical methods of decision making the preferences of decision makers have often been more or less neglected when the alternatives are evaluated (e.g., Keeney 1988, Bradshaw and Boose 1990). This has also been the case in forest management (Kangas 1992). In decision support based on operations research, the problem structuring is too often technique oriented. When developing methods based on artificial intelligence the decision theoretic methodology is typically ignored (O'Keefe 1988). Therefore, decision analyses can be markedly improved by accounting for the preferences of decision makers and other stakeholders in prioritising decision alternatives.

2. THE ANALYTIC HIERARCHY PROCESS

2.1. Basic version of the method

The Analytic Hierarchy Process (AHP), originally developed by Saaty (1977,1980), is a mathematical method for analysing complex decision problems with multiple criteria. Schmoldt et al. (2001) provided a collection of its applications and extensions in the field of forest and environmental decision support (for a review of forestry applications, see also Kangas 1999).

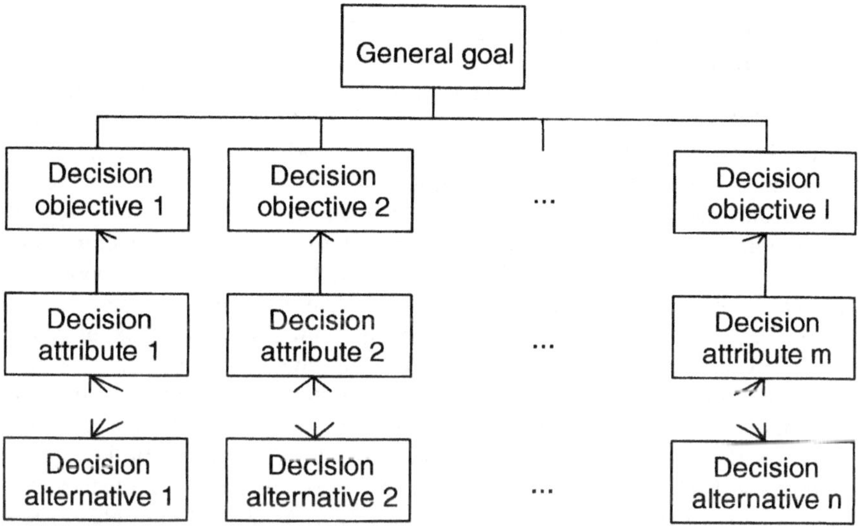

Figure 1. The basic AHP hierarchy.

Basically the AHP is a general theory of measurement based on some mathematical and psychological principles. In the method, a hierarchical decision schema is constructed by decomposing the decision problem in question into decision elements - goals, objectives, attributes and decision alternatives. The general goal is at the top of a decision hierarchy, and decision alternatives constitute the lowest level (Fig. 1). The importance or preferences of decision elements are compared in a pairwise manner with regard to the element above in the hierarchy. Based on these comparisons, an additive model on a ratio scale describing the preferences of the decision maker is estimated. The model is called a priority function. The decision alternative which produces the greatest global priority is considered the "best" and most satisfactory.

Differences in measurement scales and units do not present any difficulty when the AHP is used because the method is based on straight comparison between the significance and preference of each pair of decision elements without using any physical unit. Thus, AHP can deal with quantitative as well as qualitative attributes.

The four basic steps involved in using the AHP to address decision problems are: (1) The decision hierarchy is constructed by decomposing the original decision problem into a hierarchy of interrelated decision elements. (2) Pairwise comparisons are made at each level of the hierarchy. In making the comparison, the question is which of the two factors has a greater weight in decision making, and how much greater; or which of the two decision alternatives is more preferred with regard to a certain decision attribute. (3) Using the pairwise comparisons as input, the relative weights (importance/preference) of elements at each level are computed using the eigenvalue method. The resulting weights or priorities represent the decision-maker's perception of the relative importance or preference of the elements at each level of the hierarchy. (4) The ratings for the decision alternatives are calculated based on the relative weights of the decision elements.

Pairwise comparisons give the decision maker a basis on which to reveal his preferences by comparing two elements at a time. In the standard method presented by Saaty (1977, 1980), the decision maker has the option of expressing preferences between the two elements as: (i) equal importance or preference of both elements, (ii) weak importance or preference of one element over another, (iii) essential or strong importance or preference of one element over another, (iv) demonstrated importance or preference of one element over another, (v) absolute importance or preference of one element over another. These preferences are then translated into pairwise weights of 1, 3, 5, 7 and 9, respectively, with 2, 4, 6 and 8 as intermediate values. Many variations of the scale have been presented (see Leskinen 2001). It is also possible to carry out comparisons by using a continuous scale, e.g. by making use of graphical bars in the computer interface (e.g., Pukkala and Kangas 1993).

In step 3, the matrix of pairwise comparisons **A** is constructed. In the matrix the element $a_{ij} = 1/a_{ji}$. Thus, when $i = j$, $a_{ij} = 1$. Assuming no inconsistencies in judgements, matrix **A** has unit rank since every row is a constant multiple of the first row, and all the eigenvalues of the matrix are zero except one. Based on a consistent matrix **A**, relative weights can be determined by solving the equation

$$\mathbf{A}\mathbf{w} = \lambda\mathbf{w},$$

where λ is the only nonzero eigenvalue of a consistent matrix **A** and **w** is its right eigenvector. The solution **w** of this problem is any column of **A**. These solutions differ by a multiplicative constant. Thus, the same relative weights are obtained based on any solution. In human decision making some inconsistencies can be expected; people's feelings and preferences are not always consistent. When **A** contains inconsistencies the estimated weights can be obtained using the eigenvector equation.

$$(\mathbf{A} - \lambda_{max}\mathbf{I})\mathbf{q} = 0,$$

where λ_{max} is the largest eigenvector of matrix **A**, **q** its right eigenvector and **I** the identity matrix. The right eigenvector, **q**, constitutes the estimation of relative weights. It is the first principal component of the matrix of pairwise comparisons. If the matrix does not include any inconsistencies, i.e., the judgements made by a decision maker have been consistent, **q** is an exact estimate of the priority vector. Matrix **A** has to be estimated for each decision element at levels 1 – (n-1). Each eigenvector is scaled to sum to one to get the priorities. The form of priority functions is the same as the form of additive utility functions without interactive terms. For more details on the AHP theory and the estimation of the relative weights, readers are referred to Saaty (1977, 1980).

Global priorities of decision elements are calculated downwards from the top of the hierarchy by multiplying the local priorities by the priority of their corresponding decision element at the level above. Global priority of an element is then used to weigh the local priorities of elements at the level below and so on down to the bottom level. The sum of global priorities at each level is equal to one.

Saaty (1977) has shown that λ_{max} of a reciprocal matrix **A** is always greater or equal to n (= number of rows = number of columns). If the pairwise comparisons do not include any inconsistencies, $\lambda_{max} = n$. The more consistent the comparisons are, the closer the value of computed λ_{max} is to n. Based on this property, a consistency index, CI, has been constructed.

$$CI = (\lambda_{max} - n)/(n-1)$$

CI estimates the level of consistency with respect to the entire comparison process. A consistency ratio, CR, also measures the coherence of the pairwise comparisons. To estimate the CR, the average consistency index of randomly generated comparisons, ACI, has to be calculated ($CR = 100(CI/ACI)$). ACI varies as a function of the size of matrix (e.g. Saaty 1980). As a rule of thumb, a CR value of 10 % or less is considered to be acceptable. Otherwise, all or some of the comparisons must be repeated in order to resolve the inconsistencies of pairwise comparisons.

The Analytic Network Process (ANP) is an extension of the AHP (Saaty 2001). Basically, the ANP is a general theory of ratio scale measurement of influence with a methodology that deals with dependence and feedback. In using it one can avoid the assumption of independence among criteria in the standard AHP. The ANP consists of a so-called control hierarchy, sub-criteria controlling the interactions among competing criteria when ranking alternatives, and a network of influences among the elements and clusters in the decision structure. The ANP utilises supermatrix calculation in order to deal with interactions among the network of criteria and decision alternatives. Saaty (2001) stated that, taken generally, the ANP is more objective and more likely to capture what happens in the real world than the AHP. However, applying the ANP is much more laborious and time-consuming. Obviously, the ANP has potential application in forest management that commonly involves various interdependencies between decision elements.

Although the AHP has been a very popular tool in multi-criteria decision making, many authors have been critical of the original methodology. Moreover, some development needs have been raised in practical applications regarding the use of the standard AHP. Perhaps the two foremost problems have been that the original comparison scale does not allow for expressing any hesitation regarding the comparisons, and that the AHP itself does not provide tools for in-depth analyses of the comparisons, particularly of the uncertainty inherent in the data. The use of the ANP cannot solve these problems. However, the basic idea of performing pairwise comparisons, being a pedagogical and intuitive approach, has proved to be practical.

De Jong (1984) and Crawford and Williams (1985) showed how the pairwise comparisons data can be analysed using a regression model instead of the Saaty's eigenvalue technique. In many cases, the two methods give similar numerical results, but one major difference is that the regression model allows the analysis of uncertainties in the evaluation process. Alho and Kangas (1997) proposed a Bayesian approach to the regression model, which enables the computation of the probabilities for events such as "Decision alternative A is better than B with respect to multiple criteria". A similar system can be set up for all other regression models needed in the multi-criteria analysis. Another way to incorporate the uncertainties to the AHP calculations is to utilise fuzzy set theory (e.g. Leung and Cao 2000, Mendoza and Prabhu 2001). Then, the ratio given by the decision maker is a fuzzy number described by a membership function.

Besides multi-criteria analysis, the basic regression model can be extended in many other ways. Alho et al. (1996) extended the model to the case of multiple judges, and introduced a variance components model for the analysis of uncertainties in the evaluations. In particular, they divided the variation into three components: inter-individual variation, inconsistency shared by all judges, and residual inconsistency specific to one judge. Leskinen and Kangas (1998) showed how to analyse interval judgements instead of judgements given as a single number in the Bayesian regression framework. For example, interval judgements can be elicited in a sequence of two questions. The first question asks for a relative preference, and the second for a subjective quantification of the uncertainty of the first response. Alho et al. (2001) showed how the regression approach enables the estimation of the priorities based on incomplete pairwise comparisons matrices, i.e. on fewer comparisons.

2.2. A'WOT: using the AHP in the SWOT framework

In the so-called A'WOT method (Kurttila et al. 2000, Pesonen et al. 2001a) the Analytic Hierarchy Process (AHP) and its eigenvalue calculation framework are integrated with the SWOT analysis. The aim in applying the hybrid method is to improve the quantitative information basis of strategic planning processes. The AHP – SWOT link yields analytically determined priorities for the factors included in SWOT analysis and makes them commensurable. In addition, decision alternatives can be evaluated with respect to each SWOT factor by applying the AHP. SWOT, then, provides the basic frame within which to perform an analysis of the decision situation, and the AHP assists in carrying out SWOT more analytically and in elaborating the analysis so that alternative strategic decisions can be prioritised (Fig. 2).

SWOT is a widely applied tool in strategic decision support. In SWOT, the internal and external factors most important for the enterprise's future are grouped into four categories: Strengths, Weaknesses, Opportunities, and Threats. By applying SWOT in a strategic planning process, the aim is usually to develop and adopt a strategy resulting in a good fit between these internal and external factors.

When used properly, SWOT can provide a good basis for strategy formulation. However, SWOT could be used more efficiently than normally has been the case in its applications. The most crucial problem with SWOT is that it includes no means of analytically determining the importance of factors or of assessing the fit between SWOT factors and decision alternatives.

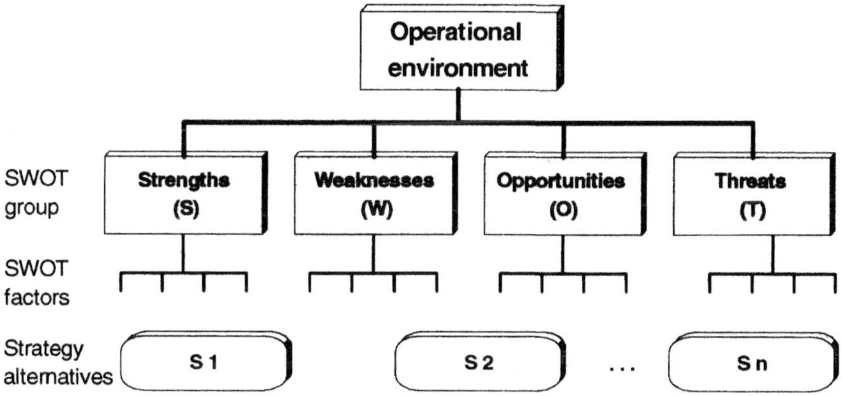

Figure 2. The A'WOT approach in terms of an AHP hierarchy (Kangas 2001b).

The main phases of A'WOT are as follows: (1) The SWOT analysis is carried out. The relevant factors of the external and internal environment are identified and included in SWOT analysis. (2) Pairwise comparisons between the SWOT factors are carried out separately within each SWOT group. When making the comparisons, the issue at stake is which of the two factors compared is more important and how much more important. With these comparisons as the input, the mutual priorities of the factors are computed. (3) The mutual importances of the SWOT groups are determined by applying pairwise comparisons. There are several methods for doing this. (4) The strategy alternatives are evaluated with respect to each SWOT factor by using pairwise comparisons and the eigenvalue technique. (5) Global priorities are calculated for the strategy alternatives.

In the earliest A'WOT applications (Kurttila et al. 2000, Pesonen et al. 2001a), only steps (1), (2), and (3), as listed above, were carried out in an early stage of a strategic planning process. A'WOT strengthens the decision basis in such cases, where the result is only the quantification and commensuration of SWOT factors. However, the final goal of any strategic planning process is to develop and propose a strategy resulting in a good fit between internal and external factors. When steps (4) and (5) are included in the A'WOT process, the initial SWOT analysis might not always be applicable as such (Pesonen et al. 2001b). This is because the SWOT factors could have been formulated so that strategy alternatives cannot be evaluated with respect to them. This being the case, SWOT factors need some value-focused modification and fine-tuning. For A'WOT, SWOT factors should be determined by asking which are the internal and external factors of the operational environment that should be taken into account in choosing the strategy for the enterprise. Then it is possible to compare strategy alternatives with respect to strengths, weaknesses, opportunities, and threats as listed in SWOT. To take an example of the pairwise comparisons: which of the two strategy alternatives compared (when implemented) makes it possible to better exploit a certain opportunity, and how much better? According to the experiences of A'WOT applications and tests, the combined use of the AHP and SWOT analysis is a promising approach in supporting strategic decision-making processes (Kurttila et al. 2000, Pesonen et al. 2000, Hobbs et al. 2001, Pesonen et al. 2001a, 2001b).

3. OUTRANKING METHODS

3.1. Pseudo-criteria

Outranking methods serve as one alternative for approaching complex choice problems with multiple criteria and multiple participants. Outranking indicates the degree of dominance of one alternative over another (e.g., Rogers and Bruen 1998b). The outranking methods enable the utilisation of incomplete value information and, for example, judgements on ordinal measurement scale (e.g., Rogers and Bruen 1998b). The problem is formulated with a set of distinct alternatives a_i, $i=1,...,n$ and a set of decision criteria g_j, $j=1,...,p$ so that $g_j(a_i)$ represents the performance of alternative i with respect to criterion j. The values of these criteria may also contain uncertainty, which can be described as either probability distributions (random variation) or fuzzy zones (uncertainty due to ignorance etc.) (Miettinen and Salminen 1999).

The uncertainty concerning the preferences and also the criterion values is dealt with using pseudo-criteria (e.g., Vincke 1992). This means that two thresholds, namely indifference and preference thresholds, are defined. The indifference threshold for criterion j, q_j, is a difference $g(a_k)$-$g(a_l)$ beneath which the decision maker is indifferent between two management alternatives a_k and a_l. The preference threshold for criterion j, p_j, is a difference above which the decision maker strongly prefers management alternative a_k over a_l . Between these two thresholds there is a zone where the decision maker hesitates between indifference and strong preference, i.e., the zone of weak preference.

The indifference threshold can be defined either with respect to the uncertainty of the criterion values or as a threshold at which the differences become perceptible to decision makers (Rogers and Bruen 1998b). Maystre et al. (1994) defined the indifference threshold as the minimum margin of uncertainty and the preference threshold as the maximum margin of uncertainty with respect to different criteria. Thus, the preference threshold implies that there is no doubt that a certain alternative is better than the other. However, there are no right values for the thresholds.

Outranking methods have both compensatory and non-compensatory elements (e.g. Pirlot 1995). In the hesitation zone, the methods are compensatory. Outside this zone, the methods have non-compensatory elements. This applies especially to ELECTRE methods, where an additional veto threshold may be applied.

3.2. The PROMETHEE method

The PROMETHEE methods were developed in the 1980's (see Brans et al. 1986). In PROMETHEE I and II, the outranking degree $\Pi(a_k,a_l)$, describing the credibility of the outranking relation that 'alternative a_k is better than alternative a_l', for each pair of alternatives (a_k,a_l) is calculated as

$$\Pi(a_k, a_l) = \sum_{j=1}^{p} w_j F_j(a_k, a_l)$$

where $F_j(a_k,a_l)$ is the preference function and w_j are the relative importance of the criteria (scaled to add up to one in the equation). The weights can be obtained, for example, by giving scores from 1 to 7 to the criteria with 1 given to the least important criteria (Hokkanen and Salminen 1997b). However, the weights could also be obtained from pairwise comparisons as in the AHP.

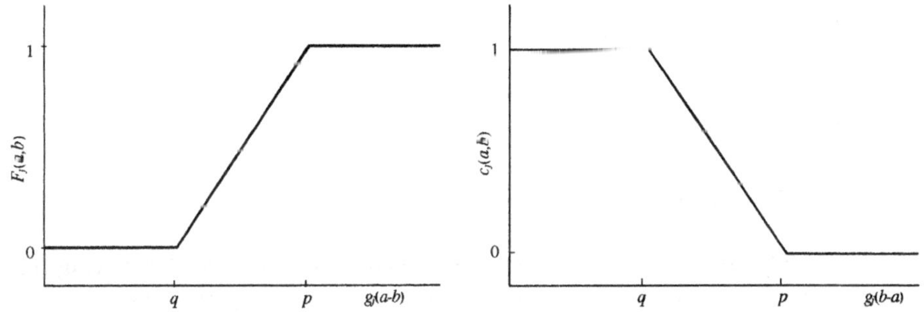

Figure 3. The preference function $F_j(a)$ *and the local concordance index* $c_j(b)$*. Note the different* x *axes (Kangas et al. 2001).*

In the PROMETHEE outranking method, the threshold values are assumed to be constant for a given criterion (see Salminen et al. 1998). The value of preference function $F_j(a_k,a_l)$ for a pair of alternatives a_k and a_l with respect to criterion j are calculated using thresholds p_j and q_j as

$$F_j(a_k,a_l)=\begin{cases}1,\text{if } g_j(a_k)-g_j(a_l)\geq p_j\\[2ex] 0,\text{if } g_j(a_k)-g_j(a_l)\leq q_j\\[2ex] \dfrac{g_j(a_k)-g_j(a_l)-q_j}{p_j-q_j}, \text{otherwise}\end{cases}$$

In this formula, a linear threshold function is utilised (Fig. 3). However, six different forms of the threshold function can be applied, which can be either linear, non-linear, or a step function (Brans et al. 1986). The criteria and threshold values together constitute the pseudo-criteria. The outranking degrees Π are used to calculate the leaving flow for each alternative,

$$\Phi^+(a_k)=\sum_{l\neq k}\Pi(a_k,a_l)/(n-1),$$

the entering flow

$$\Phi^-(a_k)=\sum_{l\neq k}\Pi(a_l,a_k)/(n-1)$$

and the net flow

$$\Phi(a_k)=\Phi^+(a_k)-\Phi^-(a_k).$$

In PROMETHEE I the alternatives are ranked based on both the leaving and entering flows. These rankings are then used to calculate a partial preorder: one alternative is deemed better than another if it is better with respect to both entering and leaving flow, otherwise the alternatives are incomparable. In PROMETHEE II the net flow is used, which leads to complete ranking (Hokkanen and Salminen 1997a,b). This ranking method utilises the 'cardinal' properties of the valuations, whereas PROMETHEE I uses the 'ordinal' properties (Bouyssou and Perny 1992, Bouyssou 1992).

3.3. The ELECTRE method

Bernard Roy (1968) originally developed the ELECTRE methods. Several versions of the ELECTRE method have been presented for different situations: ELECTRE I and IS are designed for selection problems, ELECTRE TRI for sorting problems and ELECTRE II, III and IV for ranking problems (Roy 1991, Yu 1992, Hokkanen and Salminen 1997c). ELECTRE II is an older version, where an abrupt change from indifference to strict preference is assumed instead of pseudo-criteria. The main difference between III and IV is that the relative importance indices for the different criteria are not applied in the latter.

In ELECTRE III method, a concordance index $C(a_k,a_l)$ for each pair of alternatives is computed, describing the strength of a claim that alternative a_k is at least as good as alternative a_l, utilising pseudo-criteria in a similar fashion as in PROMETHEE method. The concordance index is calculated as

$$C(a_k, a_l) = \sum_{j=1}^{p} w_j c_j(a_k, a_l)$$

where w_j are the relative importance of the criteria (scaled to add up to one in the formula) and $c_j(a_k,a_l)$ is the local concordance index, defined as (Fig. 3)

$$c_j(a_k, a_l) = \begin{cases} 0, \text{if } g_j(a_l) - g_j(a_k) \geq p_j \\ 1, \text{if } g_j(a_l) - g_j(a_k) \leq q_j \\ \dfrac{p_j - (g_j(a_l) - g_j(a_k))}{p_j - q_j}, \text{otherwise.} \end{cases}$$

In this formula, constant threshold values are applied. However, in ELECTRE III, the thresholds may be either constant, proportional to the criterion value or they could be expressed with a linear model as a function of the criterion value (e.g., Rogers and Bruen 1998b).

In ELECTRE III there is also a so-called veto threshold v_j, which is used to compute the discordance index for the alternatives. The discordance index is used to model the degree of incompensation between the criteria. This means that an alternative with a very poor value of any one criterion cannot be chosen regardless of the values of the other criteria. It also means that the weights of the criteria cannot be interpreted as substitution rates but they represent votes for the criteria (Miettinen and Salminen 1999). The discordance index is defined for each criterion as

$$d_j(a_k, a_l) = \begin{cases} 0, \text{if } g_j(a_l) - g_j(a_k) \leq p_j \\ 1, \text{if } g_j(a_l) \quad g_j(a_k) \geq v_j \\ \dfrac{(g_j(a_l) - g_j(a_k)) - p_j}{v_j - p_j}, \text{otherwise.} \end{cases}$$

The discordance indices of different criteria are not aggregated using the weights, since one discordant criterion is sufficient to discard outranking. In environmental planning the veto threshold is appropriate in the sense that some alternatives are not found acceptable at all (Rogers and Bruen 1998b). The closer the veto threshold v_j is to the preference threshold p_j, the more important the criterion j can be considered (Roy 1991).

Finally, the degree of outranking is defined by $S(a_k,a_l)$ as

$$S(a_k,a_l)=\begin{cases} C(a_k,a_l), \text{if } J(a_k,a_l)=\varnothing \\ C(a_k,a_l)\prod\limits_{j\in J(a_k,a_l)}\dfrac{1-d_j(a_k,a_l)}{1-C(a_k,a_l)}, \text{ otherwise} \end{cases}$$

where $J(a_k,a_l)$ is a set of criteria for which $d_j(a_k,a_l) > C(a_k,a_l)$ (Miettinen and Salminen 1999).

In the basic ELECTRE III method, a descending (Z_1) and ascending (Z_2) preorder is constructed using the outranking degrees S. The final partial order $Z = Z_1 \cap Z_2$ is constructed based on these two complete orders. The preorders Z_1 and Z_2 are constructed using a descending and ascending distillation. In distillation procedure, the ranking is based on the number of alternatives outranked by each alternative minus the number of alternatives that outrank it (for details, see Maystre et al. 1994).

In the obtained partial preorder some alternatives may be incomparable, i.e., their performance order cannot be determined. A complete ranking may be obtained using, for example, the 'min' procedure (Pirlot 1995). In the 'min' procedure, the alternatives are ranked according to the minimum outranking degree of each alternative. The alternative having the highest minimum is ranked first, and so on (see Miettinen and Salminen 1999). The 'min' procedure utilises the ordinal properties of the valuations (Pirlot 1995).

4. VOTING APPROACHES

4.1. Voting procedures as decision support methods

Through our democratic history, voting has proved to be an efficient tool for making choices among decision alternatives, e.g., in different kinds of elections. Nearly all people know it in one form or another. In voting theory, special attention has been paid to the systems that are neutral and difficult to manipulate (Nurmi 1987). Most voting systems are based on single criterion only or holistic preference information on decision alternatives. By its very nature, voting theory is appropriate for cases with multiple decision makers or participants, i.e., voters.

Many voting systems have a lot of common features with utility theory. The most common plurality voting, for instance, takes into account the preference ordering of the voters, even though only with respect to the best candidate. A major drawback of this method is that it is generally considered to be too easy to manipulate. It can be shown that any non-dictatorial voting scheme is subject to manipulation (Gibbard 1973, Satterthwaite 1975) but some methods require more information to manipulate an election than others (Nurmi 1987).

The approval voting method was independently proposed by several people in the 1970s (Brams and Fishburn 1983). It is a voting procedure in which each voter votes for as many candidates as she/he approves and wishes to vote for. Each candidate approved by the voter receives one vote. The candidate receiving the greatest total number of votes wins the election. This method is harder to manipulate by any voter than plurality voting. This is because it requires information about the distribution of approvals of the alternatives in order to be efficiently manipulated (Nurmi 1987). However, it tends to promote moderate candidates.

A voting procedure called the Borda count (e.g., Saari 1994) takes into account the whole preference order: given n candidates, each voter casts n votes for the most preferred candidate, n-1 votes for the second most preferred one and finally 1 vote for the least preferred candidate. A problem with this method is that the result may be a function of the number of candidates (Riker 1982).

The above-mentioned voting systems only use information about the preference ordering of the alternatives, i.e., ordinal information. So-called utilitarian voting systems also use information on the intensity of the preferences, i.e., cardinal information. These systems include, for example, direct aggregation of cardinal utilities and the demand-based method (e.g., Riker 1982, Mueller 1989). In the former, voters are required to assign utility values to the candidates, and the candidate receiving the largest sum of utility values wins. In the latter, voters offer to pay some money in order to obtain a preferred alternative, and the one getting the highest sum of bids wins. Both these methods are easy to manipulate by inflating the expressed preferences. Yet another method is the cumulative voting system, where each voter is given a number of votes they can distribute to the candidates in any way: all votes can, for instance, be given to one candidate or many candidates can be given an equal number of votes (e.g., Lakeman 1974).

Within voting theory, it has been shown that it is often in the best interest of a voter to vote for another candidate than the one most preferred by him/her, i.e., to vote strategically or to manipulate the result of the voting. Such a manipulation requires information about other voters' preferences, which is not easily available for each voter. The voters with more information have more manipulation power. In the so-called declared-strategy voting (Cranor 1996) this problem has been turned to a virtue: in this system all voters can vote strategically. Voters are first asked to give cardinal utility values for the candidates, but the utilities are not summed up like in the systems mentioned above. Then, voters submit their strategy to a computer. Such strategies may be, for example, to vote for the most probable winner in order to show support (so-called bandwagon strategy), or to vote against the least preferred alternative. These strategies are then aggregated to determine the election results. This system has been seen as especially suitable for group decision making. However, it requires special software for forming the voting strategies and aggregating them. Furthermore, the result may depend on the order in which the strategies of different voters is handled.

4.2. Multicriteria Approval

The ideas of approval voting can be utilised also in multi-criteria decision support. It can be recognised that there is a similarity between the voters as considered in the social choice theory and the criteria in multi-criteria decision making. Fraser and Hauge (1998) presented an application of approval voting, called Multicriteria Approval, which can be used in multi-criteria decision making.

In Multicriteria Approval, decision makers are only asked to rank the criteria by their importance. Then it is determined which of the alternatives will be approved for each criterion. Defining the border between approval and disapproval regarding each criterion does this. Fraser and Hauge (1998) defined the border as the average evaluation of the alternatives with respect to the criterion considered. In other words, if we have numerical criterion values $c_j(a_i)$ the approval border c_j^b is the mean criterion value $\bar{c}_j = \sum_{i=1}^{m} c_j(a_i)/m$ of the m alternatives, with respect to each criterion j. Therefore, the results of the analysis may change if new alternatives are included or some alternatives are left out.

The average value has been chosen as the borderline because, in approval voting theory, it can be proved that the optimal strategy for a voter is to vote for all the candidates that are above average (e.g., Kim and Roush 1980). The border between approval and disapproval could, however, be defined in additional ways. For example, if the utility achieved is a non-linear function of a criterion variable, the optimal borderline should be the average utility value, not the average criterion value (Fraser and Hauge 1998). The borderlines could also be defined as, for example, some proportion of the maximum value for each criterion or as some percentile of the range of variation. The borderline could also be defined in a different way for each criterion. Such an approach may not be reasonable in voting theory, where all voters are defined as equally important. In decision support, however, some criteria usually are more important than others. Then, increasing the borderline value would emphasise the importance of a criterion; decreasing it would do the opposite. If there are qualitative criteria, the preferences of decision makers can also be solicited with a verbal scale (like excellent, good, average, passable, bad) directly applicable for decision support.

The determination of the voting result begins after defining the dichotomous preferences. There are five possible classes of voting result: *unanimous*, *majority*, *ordinally dominant*, *deadlocked* and *indeterminate*. In this article the classification is presented in the way Fraser and Hauge (1998) have described it.

The voting result is unanimous if one alternative has been approved with respect to all criteria and no other alternative has been approved with respect to all criteria. A majority result occurs when one alternative has been approved with respect to the majority of criteria that have been defined to be the most important. If one alternative has been defined to be superior on grounds of the order of criteria and the dichotomous preferences, the result is ordinally dominant. Class unanimous is a subclass of majority and both of these are subclasses of ordinally dominant.

The result is deadlocked if there are two or more alternatives that are defined to belong to subclasses unanimous, majority or ordinally dominant. These alternatives are approved with respect to the same criteria. They are also disapproved with respect to the same criteria, so it is not possible to determine one single superior alternative. The result is indeterminate if there is not enough information to determine one superior alternative. In this case, there are two or more alternatives between which it is impossible to name the best, and more information is needed in addition to the criteria order.

The first phase in determining the voting class is to find out the number of existing ordinally dominant alternatives. According to Fraser and Hauge (1998) this is done by using a variation of the ordinal deductive selection system. Alternative k is classified ordinally dominant if

$$f\left(n^*\right)_{ki} \geq 0 \ \forall n^*, \ 1 \leq n^* \leq n, \ \forall i \neq k$$

where $f\left(n^*\right)_{ki} = \sum_{j=1}^{n^*} g_{ijk}$

and (assuming maximisation problem)

$$g_{ijk} = \begin{cases} 1, \ if \ \ c_j(a_k) > c_j^b \wedge c_j(a_i) \leq c_j^b \\ 0, \ if \ \ c_j(a_k) > c_j^b \wedge c_j(a_i) > c_j^b \\ 0, \ if \ \ c_j(a_k) \leq c_j^b \wedge c_j(a_i) \leq c_j^b \\ -1, \ if \ \ c_j(a_k) \leq c_j^b \wedge c_j(a_i) > c_j^b \end{cases}$$

Fraser and Hauge (1998) describe the principle of this formula as follows: "Alternative k is classified ordinally dominant if for every possible value of n^*, and all other alternatives $i \neq k$, $f(n^*)_{ki}$ remains greater than or equal to zero. If at any time, for any $i \neq k$ it becomes negative, alternative k is labelled as indeterminate." If there is only one single alternative that is ordinally dominant, the next step is to examine whether it belongs in a subclass of ordinally dominant. If it does not belong to either unanimous or majority then it is classified as ordinally dominant.

The process is based on the idea that approval with respect to a more important criterion can completely compensate for disapproval with respect to a less important criterion (Fraser and Hauge 1998). After this process each alternative is labelled as either ordinally dominant or indeterminate.

The standard version of Multicriteria Approval has been developed for a single decision maker, but if the decision makers can agree upon the importance order of the criteria and the border for approval for each criterion, the standard version suits also for group decision making.

5. STOCHASTIC MULTICRITERIA ACCEPTABILITY ANALYSIS

If the decision makers cannot or do not wish to express their preferences concerning different criteria, or even provide a preference order, it is still possible to utilise some MCDS tools in decision support. In acceptability analysis, instead of considering which alternative is preferred using certain weights, the weights that support a certain alternative are considered. Charnetski and Soland (1978) presented the comparative hypervolume criterion for cases in which only partial preference information is available. Bana e Costa (1986, 1988) presented an acceptability index, which was computed as the volume proportion of the weight space that supports a certain alternative. Lahdelma et al. (1998) introduced a stochastic version of acceptability analysis (SMAA), where the criterion values were assumed stochastic.

Assume a set of n alternatives and p criteria. The utility of alternative i is described by a linear additive model as

$$u_i = \sum_{j=1}^{p} w_j x_{ij}$$

where x_{ij} denotes the performance of alternative i with respect to criterion j, and w_j is the weight of criterion j. In this model, both the weights w and criterion values x can be uncertain. The uncertainty in criterion values is represented by the joint probability function $f(x_1, x_2, \ldots, x_p)$ (later denoted simply by $f(\mathbf{x})$) of the criterion values and the distribution of weights by their joint probability function $f(w_1, \ldots, w_p)$ (later denoted simply by $f(\mathbf{w})$), in a set of feasible weights W. Each feasible weight must be non-negative and the sum of the weights must be 1.

In SMAA, a subset of weights W_i is sought which is favourable for alternative i, meaning that among these weights alternative i is ranked first. Then, the acceptability of alternative i is calculated as the proportion of the expected volume of W_i to total weight space volume as (Lahdelma et al. 1998, Miettinen et al. 1999, Lahdelma and Salminen 2001)

$$b_j = vol(W_i) / vol(W) .$$

The volumes can be calculated as a multidimensional integral over the criteria distributions and the favourable weights giving (Miettinen et al. 1999, Hokkanen et al. 2000, Lahdelma and Salminen 2001)

$$b_j = \int_X f(\mathbf{x}) \int_{W_i} f(\mathbf{w}) d\mathbf{w}\, d\mathbf{x}$$

An acceptability index of one indicates a dominant alternative (it is the best irrespective of the weights) and an index of zero or near zero indicates an inefficient alternative (irrespective of the weights some of the alternatives is always better). Therefore, the acceptability index can be used for classifying the alternatives into more or less acceptable and not acceptable. Computationally, the analysis is carried out so that random weight and criteria vectors are generated from their distributions and the acceptability index is calculated as the proportion of realisations giving the alternative the first rank (Miettinen et al. 1999).

With this formulation there are problems, for example, when two alternatives are very close to each other. In such a case they may both have a low acceptability index. Therefore, the method was further developed (SMAA-2 method) so that the acceptability index can be calculated for any rank, not just for the first rank. Then, compromise alternatives having high acceptability for second or third rank can also be considered (Lahdelma and Salminen 2001).

Other results of the SMAA analysis are the so-called central weight vectors, which are the criteria weights representing the preferences that best support the choice of each decision alternative (Lahdelma and Salminen 2001). Finally, confidence factors can be calculated. These indicate the probability of choosing any one alternative as best, if the central weight vector is used. Low values of the confidence factor indicate that there is not enough information available for an informed decision.

There is also a version, called SMAA-O, which can utilise ordinal information. In this version, ordinal data are converted into stochastic cardinal data by simulating such values (in the [0,1] interval) to the ordinal criteria that preserve the given rankings (Miettinen et al. 1999). Rietveld and Ouwersloot (1992) have earlier presented similar ideas. In practice, p-2 random numbers are generated, and ordered together with zero and one, in order to obtain one realisation of the possible underlying cardinal criterion values.

If there is no information concerning the weights, all feasible weights are considered. However, the method can also use, for example, information about the range in which the weight lies or the importance order of the criteria. The importance order can be taken into account by sorting the generated weight vectors and giving the largest value to the most important criterion, and so on. An illustrative example of the use of SMAA-O is given in Section 6.3.

The principles of SMAA can also be used in connection to decision support tools other than additive utility function. For example, the SMAA-3 utilises ELECTRE III in SMAA framework (Hokkanen et al. 1998).

6. EXAMPLES OF MCDS APPLICATIONS IN FOREST PLANNING

6.1. Public participation in strategic planning of state forests

Pykäläinen et al. (1999) used the HIPRE program developed by Hämäläinen and Lauri (1995) in an application of interactive decision analysis (IDA) on strategic planning of state-owned natural resources. HIPRE allows the use of a modified MAUT version (MAUT = Multi-Attribute Utility Theory) of the AHP, which utilises sub-utility functions in the evaluation of choice alternatives. The Finnish Forest and Park Service (FPS) governing the case study area used the HIPRE program.

Initially, four strategies following different scenarios were formulated in the Kainuu planning project. The feasibility of land use allocations and their implications for producing forest outputs was mapped out. The impacts of the strategies were measured by numerical criterion variables and were estimated through planning calculations. So called 'Basic strategy' included the current principles of land use allocation. The 'Business strategy' emphasised economic goals of the FPS in Kainuu. The 'Forest recreation' and 'Nature conservation' strategies emphasised the related goals, respectively. However, the Finnish old forest conservation programme that was going on simultaneously required that the area of conserved forests be increased, rendering some of the initial strategies no longer feasible. Two new strategies were therefore constructed: 'Mixed 1' strategy and 'Mixed 2' strategy. 'Mixed 1' was the 'Basic' strategy including the new nature conservation criteria. 'Mixed 2' strategy was a modified version of the 'Business' strategy.

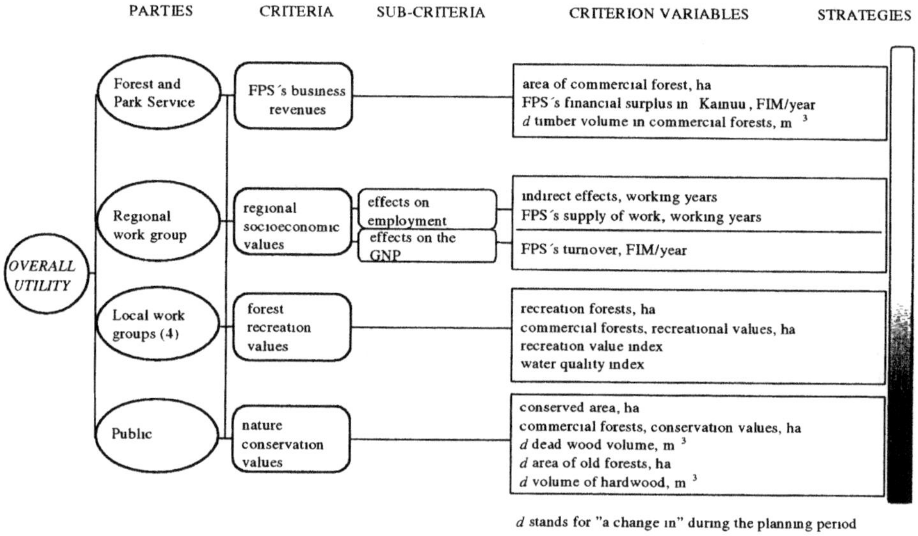

Figure 4. The decision hierarchy as applied by the Finnish Forest and Park Service in natural resources planning in Kainuu region (Pykäläinen et al. 1999).

The IDA was started with a decision hierarchy formulation (Fig. 4). The hierarchy consisted of six levels: total utility, parties, the four main criteria for forest management, the sub-criteria, the criterion variables, and the alternative forest strategies. The parties involved were FPS, one regional and four local working groups including 10 - 12 interest groups each, and the public. The main criteria were FPS's business revenues, regional socio-economic values, forest recreation values and nature conservation values.

Of these, benefits related to FPS's business revenues were measured using the area of commercial forests, FPS's financial surplus in Kainuu and the change in timber volume in commercial forests. The regional socio-economic values were measured by the effect on employment (direct and indirect) and FPS's turnover. The area of recreation forests, the area of commercial forests with recreational values, the recreation value index and the water quality index measured forest recreation values. The area of clear-cut forest stands, the area of fertilised forests and the area of ditch network maintenance measured the water quality index. The fourth main criterion was the nature conservation value, measured by the conserved area, the area of commercial forests with conservation values, the change in dead wood volume, the change in the area of old forests and the change in the volume of hardwood.

The material of this case study was reanalysed with two outranking methods, PROMETHEE II and ELECTRE III (Kangas et al. 2001). In order to produce a comparable problem formulation to the original HIPRE, the weights for the 17 criteria considered in the outranking process were those obtained with HIPRE analysis. Also, the threshold values used were related to the range of variation in the alternatives, to resemble the scaling in HIPRE.

Table 1. The public participation case ranks with three different MCDS methods. In the first ELECTRE III analysis the veto threshold was not applied and in the other analysis it was applied when the worst alternative was compared to the best (see Kangas et al. 2001).

Strategy	*HIPRE*	*PROMETHEE II*	*ELECTRE III without veto*	*ELECTRE III with veto*
Business	1	1	2	6
Basic	2	3	1	1
Forest recreation	3	2	3	3
Mixed 2	4	4	4	4
Mixed 1	5	5	5	2
Nature conservation	6	6	6	5

In PROMETHEE analysis, the most preferred alternative was the business strategy, as it was in the original HIPRE analysis (Table 1). The only difference between these methods was that the forest recreation and basic strategies changed places. The results of ELECTRE, however, differed from the other methods. If veto threshold was not utilised, the order was quite similar to the initial analysis. A veto threshold, defined as the difference between the worst and best alternative with respect to each criterion was also tested. When such a veto threshold was applied, the rank of the business strategy fell to the last place, and the compromise strategies, e.g., the Mixed 1 strategy, rose. Thus, the veto threshold is a very powerful tool.

6.2. Group decision making case in a forest holding owned by a union

The AHP, PROMETHEE I and II methods and Multicriteria Approval method were tested in a group decision making situation (Laukkanen et al. 2002). The problem was to choose a forest plan for a forest holding owned jointly by three persons, each having different objectives for forest utilisation. One owner had the following objectives: timber production, beauty of forest landscape and berry yield. The objectives of the second owner included timber production, biodiversity of the forest, beauty of forest landscape, and saving habitats for capercaillie (*Tetrao urogallus*), a valued game bird. The third owner mainly preferred timber production. All the owners had a 1/3 share of the forest holding.

The forest holding covers 30 hectares consisting of three separate blocks. The holding was divided into 13 forest stands. The planning period consisted of two 10 years periods, for which 20 alternative forest plans were produced. The values of the timber production criteria and the blueberry yield estimate were obtained using mathematical models from the Monsu planning package (Pukkala 1993). Priorities of alternative plans with respect to biodiversity considerations and habitat needs of capercaillie were based on expert judgements. The capercaillie criterion was excluded from the final analysis because it was noted that the biodiversity criterion corresponded closely enough with the capercaillie criterion. This exclusion was taken into account when defining the weight for the biodiversity. Landscape beauty was assessed using subjective weights for the predicted landscape beauty indices of the three blocks (Table 2).

In the case study the preference information was generated from the three owners in two different ways. Pairwise comparisons between the criteria were applied for AHP analysis. For PROMETHEE analysis, all decision makers defined the weights that describe the importance of the criteria using values 1–7. For Multicriteria Approval the importance order of the criteria was obtained from PROMETHEE inquiry on the basis of the average values of the decision makers' weightings. All decision makers approved this procedure. The order from the most important criterion to the least important was: net incomes, biodiversity, monetary value of the future timber production, scenic beauty of forest landscape, and berry yield. As the approval border, the average criterion value of the alternatives was used.

Table 2. Criterion values in alternative plans prepared for group decision making (Laukkanen et al. 2002).

Plan	*Net incomes, present value 2001-2020, FIM*	*Biodiversity index in 2020*	*Value of the future timber production in 2020, FIM*	*Scenic beauty index in 2020*	*Bilberry yield in 2020, kg/ha*
1	39740	1.143	876516	6.2356	10.7
2	15011	1.436	923254	6.2982	11.6
3	96900	0.419	739278	6.0517	8.2
4	151503	0.121	543862	5.7230	5.2
5	101652	0.407	752080	5.8641	8.0
6	45092	1.114	862810	6.2196	10.6
7	73225	0.441	777945	6.1927	8.7
8	134780	0.139	626711	5.3174	6.0
9	66079	0.704	826339	6.1106	9.6
10	154163	0.116	532995	5.3639	5.1
11	86666	0.471	796026	5.9070	8.7
12	133135	0.168	650071	5.7391	6.0
13	36664	1.107	883552	6.2356	10.8
14	90931	0.275	724857	6.1302	7.3
15	81864	0.543	787786	6.0517	8.9
16	114550	0.289	682128	5.9892	6.8
17	122782	0.160	577407	6.0356	5.9
18	62228	0.444	810837	6.2356	9.1
19	105142	0.232	684919	6.0517	6.7
20	95177	0.271	714019	6.1142	7.3

A different result was obtained with each of these decision support tools (Table 3). Using AHP the two alternatives with the highest incomes got the best rankings. With PROMETHEE II, the alternative with the lowest incomes and highest values for other criteria was chosen, implying less weight given to the net income than in the AHP analysis. With PROMETHEE I, seven alternatives were selected as the best and they were mutually incomparable (Fig. 5). They included the seven best alternatives of PROMETHEE II analysis. Multicriteria Approval chose one compromise solution. The results indicated that the decision models used in these methods are different but also that the methods utilise different amounts of information, which may also be of different quality. The analyses also indicated that there were several nearly equally good alternatives.

Table 3. Rankings of the 20 alternative forest plans according to different methods in the group decision-making example (see Laukkanen et al. 2002).

Alternative	*AHP*	*Promethee I*	*Promethee II*	*Promethee II with AHP weights*	*Multicriteria Approval*
1	5	1	2	7	
2	4	1	1	4	
3	11	1	7	3	1
4	1	8	18	10	
5	8	15	14	12	
6	3	1	4	5	
7	19	8	8	19	
8	7	15	19	20	
9	13	1	5	9	
10	2	15	20	18	
11	14	11	10	14	
12	6	12	17	2	
13	9	1	3	8	
14	18	15	12	16	
15	12	1	6	6	
16	10	12	13	1	
17	15	12	16	11	
18	20	8	9	20	
19	16	15	15	13	
20	17	15	11	15	

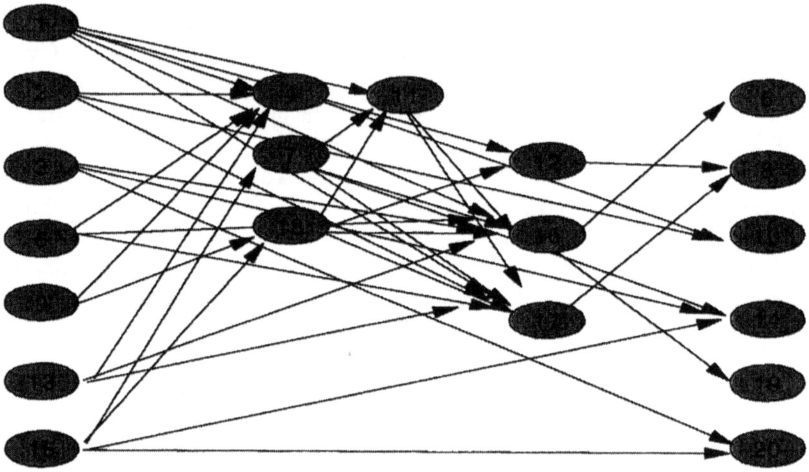

Figure 5. The partial ranking of alternatives from PROMETHEE I

6.3. SMAA-O in forest planning with both ordinal and cardinal criteria

Kangas et al. (2002) tested the SMAA-O method in supporting the choice of a tactical forest plan for a forest estate of 2024 hectares, owned by the State and administered by the Finnish Forest Research Institute (FFRI). The planning horizon was ten years (1998-2007). The forest planning case study in Kangas et al. (2000) was utilised and elaborated in the test of this method. Five criterion variables were included:

1. net incomes from timber production,
2. volume of the forest after the planning period outside the landscape ecological spots,
3. recreational value,
4. viability of rare species, and
5. risks to the watercourses.

Of these, the first two criteria could be expressed on the ratio scale while the others were on an ordinal scale (Table 4).

With respect to recreational values, forest plan alternatives were evaluated by applying an ordinal scale, where 0 = recreational value increases during the planning period, 1 = does not change to any significant extent, 2 = decreases considerably, and 3 = decreases strongly. A similar ordinal scale was applied to the criterion 'risks to the watercourse'. In this scale 0 = no risk, 1 = a considerable risk for watercourses within the area, and 2 = a considerable risk also for the water quality of the nearby lake. Evaluation of the alternative management plans with respect to the viability of rare species, in turn, was carried out using a ranking procedure. Alternatives most completely maintaining the viability of threatened and other rare species potentially persisting in the area were ranked best. Symbol 1 indicates the best ranking, 10 the worst. The net incomes were assumed to include +/- 30% uncertainty due to uncertainty in future timber prices and in timber volume prediction. The final growing stock volume was assumed to be even more uncertain, having +/- 40% uncertainty, due to uncertainty in growth and yield predictions and in the current timber volume. The uncertainty was generated from uniform distribution, assuming the errors to be mutually independent.

Table 4. Values of the alternative plans for the five criteria considered, as applied in the SMAA-O calculations (Kangas et al. 2002). For the last three criteria, low score implies good performance. Conserved volume is the growing stock volume outside conserved areas at the end of the period.

Plan number	*Net income from timber production, mill. FIM*	*Conserved volume, 1000 m^3*	*Recreation (scale 0-3)*	*Viability of rare species, ranking*	*Risks to water-course (scale 0-2)*
I	6.68	133	3	10	1
II	4.04	154	2	8	1
III	5.81	123	3	9	1
IV	3.24	143	2	5	0
V	5.52	109	2	7	1
VI	2.58	132	1	4	0
VII	4.78	106	2	6	1
VIII	2.14	125	1	2	0
IX	4.00	100	2	3	0
X	1.41	118	1	1	0

Table 5. Some results (in %) of the basic SMAA-O analysis conducted without information concerning the importance of criteria. a^h is the holistic acceptability index and r1,.., r10 are rank acceptability indices. Alternatives are sorted according to their holistic acceptability index. (Kangas et al. 2002).

Plan	a^h	*r1*	*r2*	*r3*	*r4*	*r5*	*r6*	*r7*	*r8*	*r9*	*r10*
VIII	51	21	24	15	10	8	6	5	5	4	2
VI	48	18	17	21	15	9	6	5	4	3	1
X	46	19	18	15	10	8	6	6	5	5	8
IV	41	14	12	12	15	22	9	7	5	3	1
IX	31	5	7	13	20	18	10	9	8	6	5
II	28	9	7	7	7	8	14	13	18	10	7
I	22	8	6	5	5	5	8	9	11	19	24
V	19	3	4	5	6	8	15	20	17	12	10
VII	16	1	2	4	5	8	19	18	16	14	13
III	15	3	4	5	5	5	7	9	11	24	29

Five different ecological networks were defined for the area as a starting point for elaborating alternative plans. Two wood production programmes were produced for each ecological network alternative. Thus, altogether ten alternative plans were produced. For forest stands belonging to the ecological network no treatments were proposed in any plan.

In the basic SMAA-O analysis (referred to as Analysis 1) no definite weights were determined for the criteria (Table 5). One reason for this was to get an initial insight into the method. The basic analysis served as a starting point in an interactive decision analysis process. It is widely recognised that it is hard to assign precisely "correct" weights to the criteria in any multi-criteria decision process. For that reason, it is important to study effects of changing the weights within the range of "potential" importance orders in the ranking of decision alternatives.

Three additional analyses, each with different preference information on the importance of criteria, were carried out. In Analysis 2, the greatest importance was given to the ecological criterion (4); timber-production-oriented ones (1 and 2) were seen less important than criterion 4; recreational value (3) was determined as more important than risks to watercourses (5) (Table 6). After studying the results in detail it was felt that ecological considerations were given too much weight in Analysis 2. Thus, an interactive analysis step was taken by redefining the importance order and then reanalysing choice alternatives. In Analysis 3, criteria 1, 2, and 4 were considered more important than criterion 3, criterion 5 being the least important. This was thought to reflect a more sustainable balance between timber production criteria and ecological considerations. Finally, for the sake of sensitivity analysis and of learning about trade-offs within the decision situation, Analysis 4 was performed with net incomes as the most important criterion.

Table 6. The results of SMAA-O analyses 2, 3 and 4, which used different order of importance for the criteria. Alternatives are ranked from best to worst by their holistic acceptability index. ? indicates ignorance about the importance order between criteria. (Kangas et al. 2001c).

Analysis	*Importance order of criteria*	*Ranking of plans*
1	1 ? 2 ? 3 ? 4 ? 5	VIII, VI, X, IV, IX, II, I, V, VII, III
2	4 > 1 ? 2 > 3 > 5	VIII, X, VI, IV, IX, II, VII, V, I, III
3	1 ? 2 ? 4 > 3 > 5	IV, VIII, II, VI, I, X, IX, V, III, VII
4	1 > 2 ? 4 > 3 ? 5	I, II, IV, III, V, IX, VI, VIII, VII, X

7. PERSPECTIVES ON FORESTRY MCDS

Research focusing on multiple criteria decision support has been active in recent years. In addition, MCDS methods have gained interest not only of researchers in different disciplines, but also of decision makers and planners outside the scientific community. In the forestry practice, MCDS methods have been applied in order to better meet the many challenges of today's forest management. These challenges include providing decision support for multiple-use forestry and participatory planning. It is evident that the need and demand for MCDS applications in forestry and environmental planning will further increase in the future (e.g., Tarp and Helles 1995, Kangas et al. 2000, Schmoldt et al. 2001).

The MCDS methods discussed in this paper are best suited to dealing with decision problems involving a limited number of choice alternatives. Thus, they fit with the needs of typical strategic level decision support. In tactical forest management planning, involving a huge number of decision alternatives, more efficient optimisation procedures are needed. For the purposes of tactical planning, applications of various mathematical and heuristic optimisation methods have been presented. The Data Envelopment Analysis (DEA) (see Cooper et al. 2000) is another technique that may serve the needs of tactical planning better than the MCDS methods discussed above. The HERO heuristic optimisation method (Pukkala and Kangas, 1993, Kangas et al., 2001c) especially developed for multi-objective forest planning at the forest holding or landscape level, represents hybrid approaches making use of both numerical optimisation and MAUT techniques.

Whatever the planning level and the decision support method, interactive use of the method greatly improves the efficiency of the process as well as its understandability. In practical applications, the interpretations of the parameters used in the choice models should be clear to decision makers and all participants. This is necessary in order to produce priorities or rankings that accurately reflect the preferences. In any MCDS situation there is uncertainty not only in the values of the criterion variables but also, for example, the weights of the criteria. In practical applications, a sensitivity analysis with respect to the uncertain parameters used in the calculations is thus essential. In some cases, results may show that not enough information is available for an advised decision.

There are many MCDS methods, extensions and modifications for supporting decision making in addition to those presented above. The Simple Multi-Attribute Rating Technique (SMART), originally developed by Edwards (1971), is a multi-attribute extension of the direct rating technique. Several variations of SMART have been developed (see Winterfeldt and Edwards 1986). In fact, SMART nowadays comprises a collection of different techniques and modifications instead of a single procedure. Typical for all SMART techniques is their reliance on direct numerical rating methods. The original SMART uses an additive model, but non-additive approaches have also been studied (e.g., Barzilai and Lootsma 1997). More recent developments within the SMART "family" include two approximate methods for multi-attribute utility measurement, called SMARTS and SMARTER (Edwards and Barron 1994). SMARTS utilises linear approximations to single-dimension utility functions, swing weights, and an additive aggregation model. SMARTER is a simplified version of SMARTS, and is easier to apply. In the natural resources context Reynolds (2001), among others, has applied the SMART techniques.

Several outranking-based methods have also been presented in the decision analysis literature. In addition to those already described in this chapter these include MELCHIOR (Leclercq 1984), ORESTE (Roubens 1980), REGIME (Hinloopen and Nijkamp 1982), NAIADE (Munda 1995), EXPROM-2 (Raju and Pillai 1999b), and STOPROM-2 (Raju and Pillai 1999a). Many of them include either stochastic extensions of the original methods (e.g., STOPROM-2; see also D'Avignon and Vincke 1988) or fuzzy elements (e.g., NAIADE; see also Goumas and Lygerou

2000). However, they have not gained wide application in the field of forestry or other natural resources management so far.

A method called Even Swaps developed by Hammond et al. (1998) has been used for strategy selection in a rural enterprise having forestry as one of multiple businesses (Kajanus et al. 2001). Even Swaps resembles outranking. It is based on making trade-offs among the set of criteria across a range of alternatives. The advantages and disadvantages of alternatives are systematically evened out until the best alternative can be identified. Generally taken, the approach is intuitively easy and calculation techniques in it are simple enough for nearly anyone to understand. Perhaps the main drawback of the method is the difficulty of making numerical trade-offs between measures having different units so that the trade-offs reflect one's real preferences: thinking about the value of one objective in terms of another might be hard for decision makers.

Different MCDS methods often give somewhat different rankings for the alternatives. However, the methods and the results are not necessarily directly comparable as such although they have been applied to the same planning case. The inconsistencies in the results yielded by different methods might be due to the fact that 1) the choice problem formulations do not reflect the same preference structures and 2) the ways in which preference information is processed vary between different methods. Also, the methods interpret the model parameters (most importantly the criterion weights) differently. From the application point of view, it may be more important to compare the qualities of the methods themselves –especially how they fit with different practical decision making situations (Table 7)– than to compare the results obtained by different methods.

Table 7. Comparison of some MCDS methods concerning their preference input needs and the characteristic of the results provided by them.

Method	*Information from decision maker*	*Result*
AHP	Pairwise comparison of decision elements (goals, alternatives, etc.)	Cardinal priorities for the alternatives
SMART	Basic version: rating of criteria compared to the most or least important one	Cardinal priorities for the alternatives
Outranking	Criteria weights (and possibly threshold values)	Partial ranking (e.g. PROMETHEE I and ELECTRE III) – complete ranking (e.g. PROMETHEE II)
Multicriteria Approval	Importance order of criteria	Dominant alternative(s)
SMAA	Anything that the stakeholders provide	Acceptability of alternatives

For example, the ability to deal with uncertain and fuzzy information is an indisputable advantage of the outranking methods. A further advantage of outranking is that, when compared to MAUT approaches, the preference estimation procedures are versatile and diverse. For example, the veto option enables the user to express constraints pragmatically and in an intuitively reasonable way. In MAUT-based methods, multiplicative partial utility functions can serve similar purposes (e.g., Kangas et al. 2001c). However, tools acting as constraints should be applied with care, because they may have dramatic effects on the results of the analysis.

A clear disadvantage of the outranking methods, when compared to the AHP, for instance, is that the methodology used in the analyses is complicated and hard to explain to non-specialists, although the method is very easy to use. Transparency of the method and the interpretability of the results are important qualities of planning methods, particularly in participatory planning. Indeed, the situation being the opposite constitutes perhaps the most crucial shortcoming of most outranking methods from the viewpoint of practical applications.

Of the methods presented in this paper, Multicriteria Approval has had only a few applications in any field so far. This is understandable since the method is quite new. However, it seems to have much potential for natural resources decision support. The method is suitable in situations where information demanded by other methods is difficult or expensive to obtain, or the available information is of low quality. It is easy to explain to the decision makers and it is understandable. There can be different kinds of decision makers, experts and non-experts, and the method is still fair for everyone. Due to minimal information requirements, costs of using this method remain low. Since Multicriteria Approval is so easy for decision makers, queries can be carried out using mail inquiry or, for example, the Internet. Approaches based on voting theory would be worth applying especially in participatory and group planning situations. However, in cases where cardinal weighting of criteria, and cardinal preference orderings of the alternatives with respect to single criteria, would be available, applying dichotomous preferences results in loss of information.

The comparison of Multicriteria Approval with other methods applied to similar decision support purposes, such as MAUT and outranking, is interesting. Outranking methods used for ranking alternatives typically produce an ordinal preference order. In some cases, like PROMETHEE II, the rankings also have cardinal properties. The MAUT methods, for instance AHP and SMART, give measures of relative priorities for the alternatives. Preference information is also expressed in different ways by different approaches. The MAUT methods are typically based on cardinal preferences. In outranking methods the weights of the criteria need to be defined but the method can use either ordinal or cardinal information. The most often used MAUT methods are based on a linear additive utility model, and thus they are fully compensatory. Outranking methods have both compensatory and non-compensatory elements. Multicriteria Approval, on the other hand, is clearly non-compensatory.

When Multicriteria Approval or outranking is applied to group or participatory decision support, a mutual rank order of the criteria is determined. In the AHP (or other cardinal utility based methods) it is possible to assess the criteria weights separately for each stakeholder and then computationally combine the preference information obtained. For combining the separate analyses, each decision maker or participant is assigned a unique weight. Using common weights or even common rank of criteria may seem theoretically problematic. A well-known theorem of Arrow (1951) states that there is no method of aggregating individual preferences over three or more alternatives that would satisfy several conditions for fairness and always produce a logical result. However, in any practical group decision support the preferences of the stakeholders need to be combined in one way or another, otherwise the only decision support left would be negotiation. Yet, negotiations may be even more susceptible to manipulation than any computational decision support techniques.

The SMAA family can utilise many different preference structures and problem formulations. In principle, any decision support tool could be utilised in the SMAA context. Also, no definite information requirements exist – in fact the SMAA methods can be used in cases of extremely low quality information as well as very high quality information, even in the same problem. Also the possibilities for expressing the uncertainties involved are manifold. However, SMAA methods do not provide any clear recommendation for action, as the other methods do. This characteristic could be either a drawback or a benefit, depending on the situation.

Some researchers have seen different methods yielding different results to be a major disadvantage of the MCDS approach (e.g., Gershon and Duckstein 1983, Voogd 1983, see also Zanakis et al. 1998). We do not find this to be a serious problem. Forestry planning problems are often very complex and involve much uncertainty. In addition, any environmental decision support must be performed using incomplete information. Different methods allow different problem formulations, and different calculation procedures. No single method can be expected to tell the one and only truth concerning any natural resources planning problem but each method may give some additional decision support. According to many comparative studies (e.g., Karni et al. 1990, Lootsma 1990, Goicoechea et al. 1992), different methods give quite similar results in real-life problems – provided that the methods reflect the same preference information. However, the differences between results obtained by the different methods may be profound if this prerequisite does not hold, the differences between the alternatives are small and there are many alternatives to be compared (Laukkanen et al. 2002).

In most situations, it seems appropriate to utilise several decision aid methods in dealing with the same problem if the choice of the method is not self-evident. In that case, if the methods do not agree, the decision makers could be given the solutions from different methods with an explanation as to why they differ. Then the decision makers can choose one of these alternative solutions (Salminen et al. 1998), guided by a consultant.

It is always hard to adopt a new way of approaching choice problems and to learn to use and understand new methods. Hybrid methods, if they include qualities of the 'old' methods with which people are already familiar, can promote the introduction of new ideas into planning processes. Although not necessarily orthodox from the decision theoretical viewpoint, applying hybrid methods or multiple methods simultaneously to the same decision problem may serve the purpose from the behavioural and educational point of view, too.

In application development work it is important to pay special attention to the interpretability and comprehensibility of use of the methods and of the results they provide, in order to produce applications more useful in practical forest planning. Another interesting topic of future research would be to study the choice of the most suitable decision support method(s) in different kinds of planning situations within the forestry context (see, e.g., Ozernoy 1992). The behavioural viewpoint in applying MCDS methods in practical tasks is very important. The methods applied should be approved by the decision makers and participants. In addition, they should support education and learning about the planning situation at hand, and about various aspects of decision making in general.

REFERENCES

Alho, J.M. and Kangas, J. 1997. Analysing uncertainties in experts' opinions of forest plan performance. *Forest Science* 43: 521-528.

Alho, J.M., Kangas, J. and Kolehmainen, O. 1996. Uncertainty in the expert predictions of the ecological consequences of forest plans. *Applied Statistics* 45: 1-14.

Alho, J.M., Kolehmainen, O. and Leskinen, P. 2001. Regression methods for pairwise data. In: Schmoldt, D., Kangas, J., Mendoza, G.M. & Pesonen, M. (Eds.). *The Analytic Hierarchy Process in Natural Resources and Environmental Decision Making*. Kluwer Academic Publishers, Managing Forest Ecosystems 3. Pp. 235-252.

Arrow, K. 1951. *Social choice and individual values.* Wiley, New York.

Bana e Costa, C.A. 1986. A multicriteria decision aid methodology to deal with conflicting situations on the weights. *European Journal of Operational Research* 26: 22-34.

Bana e Costa, C.A. 1988. A methodology for sensitivity analysis in three criteria problems: A case study in municipal management. *European Journal of Operational Research* 33: 159-173.

Barzilai, J. and Lootsma, F.A. 1997. Power relations and group aggregation in the multiplicative AHP and SMART. *Journal of Multi-Criteria Decision Analysis* 6: 155-165.

Bouyssou, D. 1992. Ranking methods based on valued preference relations: a characterization of the net flow method. *European Journal of Operational Research* 60: 61-67.

Bouyssou, D. and Perny, P. 1992. Ranking methods for valued preference relations: a characterization of a method based on leaving and entering flows. *European Journal of Operational Research* 61: 186-194.

Bradshaw, J. M. and Boose, J. H. 1990. Decision analysis techniques for knowledge acquisition: combining information and preferences using Aquinas and Axotl. *International Journal of Man-Machine Studies* 32:121-186.

Brams, S.T. and Fishburn, P. 1983. *Approval voting*. Birkhauser, Boston.

Brans, J.P., Vincke, Ph. and Mareschal, B. 1986. How to select and how to rank projects: The PROMETHEE method. *European Journal of Operational Research* 24: 228-238.

Charnetski, J.R. & Soland, R.M. 1978. Multiple-attribute decision making with partial information: the comparative hypervolume criterion. *Naval Research Logistics Quarterly* 25: 279-288.

Cooper, W.W., Seiford, L.M. and Tone, K. 2000. Data Envelopment Analysis. A comprehensive text with models, applications, references and DEA-solver sotfware. Kluwer Academic Publishers. 318 p.

Cranor, L.F. 1996. Declared-strategy voting: an instrument for group decision making. Academic dissertation. http://www.research.att.com/~lorrie/pubs/diss/diss.html

Crawford, G. and Williams, C. 1985. A note on the analysis of subjective judgment matrices. *Journal of Mathematical Psychology* 29: 387-405.

d'Avignon, G.R. and Vincke, Ph. 1988. An outranking method under uncertainty. *European Journal of Operational Research* 36: 311-321.

De Jong, P. 1984. A statistical approach to Saaty's scaling method for priorities. *Journal of Mathematical Psychology* 28: 467-478.

Edwards, W. 1971. Social utilities. *Engineering Economist. Summer Symposium Series* 6: 119-129.

Edwards, W. and Barron, F.H. 1994. SMARTS and SMARTER: Improved simple methods for multiattribute utility measurement. *Organizational Behavior and Human Decision Processes* 60:306-325.

Forman, E. H. 1987. Relative vs. absolute worth. *Mathematical Modelling* 9(3-5):195-202.

Fraser, N.M. and Hauge, J.W. 1998. Multicriteria approval: application of approval voting concepts to MCDM problems. *Journal of Multi-Criteria Decision Analysis* 7: 263-272.

Gershon, M. E. and Duckstein, L. 1983. Multiobjective approaches to river basin planning. *Journal of Water Resource Planning* 109: 13-28.

Gibbard, A. 1973. Manipulation of voting schemes: a general result. *Econometrica* 41: 587-601.

Goumas, M. and Lygerou, V. 2000. An extension of the PROMETHEE method for decision making in fuzzy environment: ranking of alternative energy exploitation projects. *European Journal of Operational Research.* 123: 606-613.

Guariso, G. and Werthner, H. 1989. *Environmental decision support systems.* Ellis Horwood. Chichester. 240 p.

Hammond, J.S., Keeney, R.L. and Raiffa, H. 1998. Even Swaps: A rational method for making trade-offs. *Harward Business Review*, March-April. Pp. 1-11.

Hinloopen, E. and Nijkamp, P. 1982. *Information systems in an uncertain planning environment – some methods.* Working paper 82-117, International Institute for Applied Systems Analysis.

Hobbs, M., Hytönen, L. and Kangas, J. 2001. Factors affecting the economic sustainability of the non-industrial private forest enterprise: a comparison of stakeholder perceptions. In: Vayrynen, J. & Niskanen, A. (eds.). *International Symposium on Economic Sustainablity of Small-scale Forestry.* IUFRO Working Unit 3.08.00: Small-scale Forestry. 20-26 March, 2001, Joensuu, Finland. Abstracts. p. 60.

Hokkanen, J., Lahdelma, R. and Salminen, P. 2000. Multicriteria decision support in a technology competition for cleaning polluted soil in Helsinki. Journal of Environmental Management 60:339-348.

Hokkanen, J., Lahdelma, R., Miettinen, K. & Salminen, P. 1998. Determining the implementation order of a general plan by using a multicriteria method. *Journal of Multi-Criteria Decision Analysis* 7: 273-284.

Hokkanen, J. and Salminen, P. 1997a. Choosing a solid waste management system using multicriteria decision analysis. *European Journal of Operational Research* 98: 19-36.

Hokkanen, J. and Salminen, P. 1997b. Locating a waste treatment facility by multicriteria analysis. *Journal of Multi-Criteria Decision Analysis* 6: 175-184.

Hokkanen, J. and Salminen, P. 1997c. ELECTRE III and IV decision aids in an environmental problem. *Journal of Multi-Criteria Decision Analysis* 6:215-226.

Hamalainen, R. and Lauri, H. 1995. *Hipre3+ Users Guide.* Helsinki University of Technology. Systems analysis Laboratory.

Kajanus, M. 2001. *Strategy and innovation model for the entrepreneurial forest owner.* University of Joensuu, Faculty of Forestry. 44 p + 4 app. Doctoral thesis.

Kangas, A., Kangas, J. and Pykäläinen, J. 2001. Outranking methods as tools in strategic natural resources planning. *Silva Fennica* 35: 215-227.

Kangas, J. 1992. *Metsikön uudistamisketjun valinta - monitavoitteiseen hyötyteoriaan perustuva päätösanalyysimalli.* Summary: Choosing the regeneration chain in a forest stand: A decision model based on multi-attribute utility theory. University of Joensuu. Publications in Sciences 24. 230 p. Doctoral thesis.

Kangas, J. 1993. A multi-attribute preference model for evaluating the reforestation chain alternatives of a forest stand. *Forest Ecology and Management* 59: 271-288.

Kangas, J. 1999. The analytic hierarchy process (AHP): Standard version, forestry application and advances. In: Helles, F., Holten-Andersen, P. & Wichmann, L. (Eds.). *Multiple Use of Forests and Other Natural Resources.* Kluwer Academic Publishers, Forestry Sciences 61. Pp. 96-105.

Kangas, J., Store, R., Leskinen, P. and Mehtätalo, L. 2000. Improving the quality of landscape ecological forest planning by utilizing advanced decision-support tools. *Forest Ecology and Management* 132:157-171.

Kangas, J., Hytönen, L. and Loikkanen, T. 2001a. Integrating the AHP and HERO into the process of participative natural resources planning. In: Schmoldt, D., Kangas, J., Mendoza, G.M. & Pesonen, M. (Eds.). *The Analytic Hierarchy Process in Natural Resources and Environmental Decision Making.* Kluwer Academic Publishers, Managing Forest Ecosystems 3. Pp. 131-148.

Kangas, J., Pesonen, M., Kurttila, M. & Kajanus, M. 2001b. A'WOT: Integrating the AHP with SWOT analysis. In: Dellman, K. (Ed.). *Proceedings of the Sixth International Symposium on the Analytic Hierarchy Process ISAHP 2001*, August 2-4, 2001, Kursaal Bern, Switzerland. Pp. 189-198.

Kangas, J., Pukkala, T. and Kangas, A.S. 2001c. HERO: Heuristic Optimisation for Multi-Criteria Forestry Decision Analysis. In: Schmoldt, D., Kangas, J., Mendoza, G.A. & Pesonen, M. (Eds.). *The Analytic Hierarchy Process in Natural Resource and Environmental Decision Making.* Kluwer Academic Publishers, Managing Forest Ecosystems 3. Pp. 51-65.

Kangas, J., Hokkanen, J., Kangas, A., Lahdelma, J. & Salminen, P. 2002. Applying stochastic multicriteria acceptability analysis to forest ecosystem management with both cardinal and ordinal criteria. Manuscript.

Keeney, R. L. 1982. Decision analysis: An overview. *Operations Research* 30: 803-838.

Keeney, R. L. 1988. Value-driven expert systems for decision support. *Decision Support Systems* 4: 405-412.

Kim, K.H. and Roush, R.W. 1980. *Introduction to mathematical consensus theory*. New York, Marcel Dekker.

Kurttila, M., Pesonen, M., Kangas, J. and Kajanus, M. 2000. Utilizing the analytical hierarchy process (AHP) in SWOT analysis – A hybrid method and its application to a forest–certification case. *Forest Policy and Economics* 1: 41-52.

Lahdelma, R., Hokkanen, J. and Salminen, P. 1998. SMAA – Stochastic multiobjective acceptability analysis. *European Journal of Operational Research* 106:137-143.

Lahdelma, R. and Salminen, P. 2001. SMAA-2: Stochastic multicriteria acceptability analysis for group decision making. *Operations Research* 49(3): 444-454.

Lakeman, E. 1974. *How democracies vote: a study of electoral systems*. Faber & Faber, London, 4th edition.

Laukkanen, S., Kangas, A. and Kangas, J. 2002. Applying voting theory in natural resource management: a case of multiple-criteria group decision support. *Journal of Environmental Management* 64: 127-137.

Leclercq, J.P. 1984. Propositions d'extension de la notion de dominance en présence de relations d'ordre sur le pseudo-critères: MELCHIOR. *Mathematical Social Sciences* 8:45-61.

Leskinen, P. 2001. *Statistical methods for measuring preferences*. University of Joensuu, Publications in Social Sciences 48. 111 p. Doctoral thesis.

Leskinen, P. and Kangas, J. 1998. Analysing uncertainties of interval judgement data in multiple-criteria evaluation of forest plans. *Silva Fennica* 32: 363-372.

Leung, L.C. and Cao, D. 2000. On consistency and ranking of alternatives in fuzzy AHP. *European Journal of Operational Research*. 124:102-113.

Maystre, L.Y., Pictet, J. and Simos, J. 1994. *Méthodes multicritères ELECTRE*. Presses Polytechniques et Universitaires Romandes, Lausanne.

Mendoza, G.A. and Prabhu, R. 2001. A fuzzy Analytic Hierarchy Process for assessing biodiversity conservation. In: Schmoldt, D., Kangas, J., Mendoza, G.A. & Pesonen, M. (Eds.). *The Analytic Hierarchy Process in Natural Resource and Environmental Decision Making*. Kluwer Academic Publishers, Managing Forest Ecosystems 3. Pp. 219-235.

Miettinen, K. and Salminen, P. 1999. Decision-aid for discrete multiple criteria decision making problems with imprecise data. *European Journal of Operational Research* 119: 50-60.

Miettinen, K., Lahdelma, R. and Salminen, P. 1999. *SMAA-O – Stochastic multicriteria acceptability analysis with ordinal criteria*. University of Jyväskylä. Reports of the Department of Mathematical Information Technology Series B. Scientific Computing No. B 5.

Mueller, D.C. 1989. *Public choice II*. Cambridge University Press, New York.

Munda, G. 1995. *Multicriteria evaluation in a fuzzy environment*. Springer-Verlag, New York.

Nurmi, H. 1987. *Comparing voting systems*. D. Reidel Publishing Company, Dordrecht.

O'Keefe, R. M. 1988. Artificial intelligence and the management science practitioner: expert systems and MS/OR methodology (Good News and Bad). *Interfaces* 18(6):105-113.

Ozernoy, V. M. 1992. Choosing the 'best' multiple criteria decision-making method. *INFOR* 30: 159-171.

Pirlot, M. 1995. A characterization of 'Min' as a procedure for exploiting valued preference relations and related results. *Journal of Multi-Criteria Decision Analysis* 4: 37-56.

Pesonen, M., Ahola, J., Kurttila,M., Kajanus, M. and Kangas, J. 2001a. Applying A'WOT to Forest Industry Investement Strategies: Case Study of a Finnish Company in North America. In: Schmoldt, D., Kangas, J., Mendoza, G.A. & Pesonen, M. (Eds.). *The Analytic Hierarchy Process in Natural Resource and Environmental Decision Making*. Kluwer Academic Publishers, Managing Forest Ecosystems 3. Pp. 131-148.

Pesonen, M., Kurttila, M., Kangas, J., Kajanus, M. and Heinonen, P. 2001b. Assessing the priorities among resource management strategies at the Finnish Forest and Park Service. To appear in *Forest Science*.

Pukkala, T. 1993. Monikäytön suunnitteluohjelmisto MONSU. Ohjelmiston toiminta ja käyttö. Mimeograph. University of Joensuu, Faculty of Forestry. 42 p.

Pukkala, T. and Kangas, J. 1993. A heuristic optimization method for forest planning and decision-making. *Scandinavian Journal of Forest Research* 8: 560-570.

Pykäläinen, J., Kangas, J. and Loikkanen, T. 1999. Interactive decision analysis in participatory strategic forest planning: experiences from State owned boreal forests. *Journal of Forest Economics* 5: 341-364.

Raju, K.S. and Pillai, C.R.S. 1999a. Multicriteria decision making in performance evaluation of an irrigation system. *European Journal of Operational Research* 112: 479-488.

Raju, K.S. and Pillai, C.R.S. 1999b. Multicriteria decision making in river basin planning and development. *European Journal of Operational Research* 112: 249-257.

Reynolds, K.M. 2001. Prioritizing salmon habitat restoration with the AHP, SMART, and uncertain data. In: Schmoldt, D., Kangas, J., Mendoza, G.M. & Pesonen, M. (Eds.). *The Analytic Hierarchy Process in Natural Resources and Environmental Decision Making*. Kluwer Academic Publishers, Managing Forest Ecosystems 3. Pp. 199-218.

Riker, W.H. 1982. *Liberalism against populism*. Waweland Press, Inc, Prospect Heights.

Rietveld, P. and Ouwersloot, H. 1992. Ordinal data in multicriteria decision making, a stochastic dominance approach to siting nuvlear power plants. *European Journal of Operational Research* 56:249-262.

Rogers, M. and Bruen, M. 1998a. A new system for weighting environmental criteria for use within ELECTRE III. *European Journal of Operational Research* 107: 552-563.

Rogers, M. and Bruen, M. 1998b. Choosing realistic values of indifference, preference and veto thresholds for use with environmental criteria within ELECTRE. *European Journal of Operational Research* 107: 542-551.

Roubens, M. 1980. Analyse et agrégation des préférences: modélisation, ajustement et résumé de données relationnelles. *Revue Belge de Statistique, d'Informatique, et de Reserche Opérationelle* 20:36-67.

Roy, B. 1968. Classement et choix en présence de points de vue multiples (la méthode ELECTRE). *Revue Française d'Informatique et de Recherche Opérationnelle* 8:57-75.

Roy, B. 1991. The outranking approach and the foundations of Electre methods. *Theory and Decision* 31:49-73.

Saari, D.G. 1994. *Geometry of voting*, volume 3 of Studies in Economic Theory. Springer-Verlag, New York.

Saaty, T.L. 1977. A scaling method for priorities in hierarchical structures. *Journal of Mathematical Psychology* 15: 234-281.

Saaty, T. L. 1980. *The Analytic Hierarchy Process*. McGraw-Hill, New York. 287 p.

Saaty, T.L. 2001. *Decision Making With Dependance And Feedback. The Analytic Network Process*. 2nd edition. RWS Publications. 376 p.

Salminen, P., Hokkanen, J. and Lahdelma, R. 1998. Comparing multicriteria methods in the context of environmental problems. *European Journal of Operational Research* 104: 485-496.

Satterthwaite, M.A. 1975. Strategy-proofness and Arrow's conditions: Existence and correspondence theorems for voting procedures and social welfare functions. *Journal of Economic Theory* 10: 187-217.

Schmoldt, D., Kangas, J., Mendoza, G.A. and Pesonen, M. (Eds.). 2001. *The Analytic Hierarchy Process in Natural Resource and Environmental Decision Making*. Kluwer Academic Publishers, Managing Forest Ecosystems 3. 305 p.

Tarp, P. and Helles, F. 1995. Multi-criteria decision-making in forest management planning. *Journal of Forest Economics* 1: 273-306.

Vincke, Ph. 1992. *Multi-criteria decision aid*. John Wiley & Sons. 154 p.

von Winterfeldt, D. 1988. Expert systems and behavioral decision research. *Decision Support Systems* 4: 461-471.

von Winterfeldt, D. and Edwards, W. 1986. *Decision analysis and behavioral research*. Cambridge University Press. Cambridge. 604 p.

Voogd, H. 1983. *Multicriteria Evaluation for Urban and Regional Planning*; Pion, London.

Zanakis, S.H., Solomon, A., Wishart, N. and Dublish, S. 1998. Multi-attribute decision making: A simulation comparison of select methods. *European Journal of Operational Research* 107: 507-529.

Yu, W. 1992. *Aide multicritère à la décision dans le cadre de la problématique du tri: méthodes et applications*. PhD thesis, LAMSADE, Université Paris-Dauphine, Paris.

Chapter 4

MULTIDIMENSIONAL MEASUREMENTS AND APPROACHES TO FOREST SUSTAINABILITY ASSESSMENTS

An overview of models, approaches, and issues

GUILLERMO A. MENDOZA
Department of Natural Resources and Environmental Sciences, University of Illinois, Urbana, Illinois, USA

RAVI PRABHU
Center for International Forestry Research, Harare, Zimbabwe

Abstract This article provides an overview of concepts, models, measures, and dimensions of forest sustainability assessments. The methods described include models that are fully developed and have actually been applied to various application contexts, as well as two normative models, graph theory and cognitive mapping, that appear to have potential for more in-depth analysis of the interactions of sustainability indicators. The former is based on the principles of multi-criteria analysis (MCA), an approach that is capable of accommodating the amount of complexity and uncertainty inherent in the concept and practice of forest sustainability. The other methods make use of "softer" analyses that are suited particularly for the assessment and measurement of qualitative indicators of sustainability. The MCA models are capable of determining relative importance of indicators and estimating sustainability index values based on the combined impacts of indicators. Measures of relative importance can be used to prioritise indicators or filter out those that are relatively unimportant. 'Softer' analysis methods such as cognitive mapping and graph theory can help examine the interactions between and among indicators and help identify critical domains, critical paths, and tactically and strategically important indicators, based on the strength of their connectivity with other indicators, either directly or indirectly. These measures of connectivity can provide useful insights that can help managers identify indicators in need of attention, monitoring, or even mitigation when necessary. Concepts such as the use of sustainability thresholds and qualitative flags are also discussed.

Keywords: cognitive mapping, critical thresholds, flag theory, graph analysis, multi-criteria analysis, sustainability index

Timo Pukkala (ed.), Multi-objective forest planning, 71–98.

1. INTRODUCTION

Over the last few years a significant body of work has emerged about the principle of sustainability. There is now widespread consensus among scientists and practitioners on the importance of sustainability and its desirability as a goal for environmental management and for managing different resources (Pelt et al. 1995, Renning and Wiggering 1997). Sustainability has also become a major goal of many government policies (Committee of Scientists 1999) including research (NRC 1998) and corporate strategies (AF&PA 1995). The literature is rich with studies describing sustainable management of a variety of natural resources such as forests (ITTO 1992, Sample and Aplet 1994, IUFRO 1997, Mendoza and Prabhu 2000a-d), agricultural production (Neher 1992, Ikerd 1993, Hansen 1996), agroforestry (Cramb 1993, Torquebiau 1992), land management (Cornforth 1999), animal production (Spedding 1995, Vavra 1996), livestock production (Udo and Cornelissen, 1998), farming systems (Bockstaller et al. 1997) and food production (Brlacich et al. 1991).

Despite widespread agreement about the significance and merit of the sustainability concept, there is still some confusion about its meaning. Bell and Morse (1999) have raised the issue that the concept is too broad:

> Given its ubiquitous use and popularity, the lack of a concrete definition of sustainability appears to be very surprising. How can something so vague be so popular?

On the other hand, Poore (1990) has noted that a very precise definition of sustainability would be too restrictive to be broadly applicable. Thus in essence we have a broad societal consensus, as reflected in global policy documents such as the Rio Principles and various international conventions, about an inherently fuzzy concept that works because it defines a principle narrowly enough to be broadly meaningful across the globe but broadly enough to allow narrower interpretations in local situations.

Among the key themes of sustainability on which consensus has emerged are: utilization of resources without compromising their health and productivity (Costanza et al. 1992); the integration of economic, ecological, and biophysical components of resource ecosystems (Renning and Wiggering 1997, Munda 1998, Hermanidez and Nijkamp 1998); and the importance of considering the temporal and spatial dimensions of sustainability (Gray 1991, Conway 1994).

Given that sustainability will always have different local interpretations, the practitioner who wishes to assess performance of local systems against this principle in a transparent and readily comparable manner faces a considerable challenge. This paper therefore aims to provide an overview of concepts and approaches that might be used for such sustainability assessments. It describes different measures and dimensions of sustainability. Further, it provides an overview of methods and procedures, qualitative or quantitative, structured or semi-structured (soft methods), and the different analytical constructs or tools that can provide deeper understanding into the intricacies of forest sustainability assessments.

2. OPERATIONALIZING THE CONCEPT OF SUSTAINABILITY

2.1. Criteria and indicators

For sustainability to be meaningful and useful, practical management guides have to be developed to make it operational. Lal (1991) has argued that the "primary concern is not whether sustainability is important or not but how to use the concept in practical and operational terms to alleviate the pressing problems and production constraints'.

Over the last decade, there has been significant investment of resources both nationally and internationally to develop ways of implementing the concept. One of the most common initiatives is the development of indicators for assessing sustainability (Gray 1991, Heinen 1994). The term indicator has been used in a variety of ways according to general subject, context, and areas of concern. It has, for example, been used traditionally in economic applications where the economic or development status of a nation or region is assessed and monitored using socio-economic indicators (Gray 1991, Verbruggen and Kuik 1991, Renning and Wiggering 1997, Mebratu 1998). In plant and animal ecology, use of a key indicator species is a core concept in characterizing and managing plant and animal ecosystems.

In a generic sense, 'indicators' can be viewed as 'factors' or variables that can be used to measure the status and change of a system or process. Following this generic sense, indicators in general can be used as measures to evaluate sustainability. In forestry, assessment of sustainability has evolved towards the use of criteria and indicators (Prabhu et al. 1996, Lammerts and Blom 1997). Following this strategy, a number of initiatives have been proposed to define parameters for measuring sustainable forestry at different geographic scales and administrative or management levels; that is global, regional, national or specific forest management units. These initiatives have taken different forms; for example, standards that could be used for forest certification (ITTO 1992, ISO 1995, Vogt 1999), or general principles involving criteria and indicators (C&I) that could be used for the assessment of forest conditions (FSC 1994, SCS 1994, SGS 1994, Vogt et al. 1999).

Most of the criteria and indicators for sustainable forestry are holistic in the sense that they embrace a wide range of forest management issues that include economic, environmental, biological, and physical factors. Prabhu et al. (1996, 1998) typify these sets of criteria and indicators by organizing them in a hierarchical framework structured under three general theoretical constructs, namely, *Principles, criteria* and *indicators* as shown in Figure 1, p. 77. Given this hierarchical framework, a forest can be assessed through specific evaluations with respect to measured values of the criteria and indicators found relevant to the particular forest. The hierarchical structure of defining indicators allows the deconstruction of the complex sustainability assessment problem into manageable bits that can lend themselves to more structured or formal analysis.

2.2. Multi-criteria analysis (MCA) in assessing forest sustainability

The developing literature on sustainability continues to expand because of the challenges it poses. While much of the literature has critically examined the concept, little has been done in terms of providing theoretical support that can transform its appealing rhetoric into something operational and practical. This points to the need for a systematic analytical framework that embraces the multiple dimensions of the sustainability principle and its equivocal nature. Such a framework must transcend the inherent ambiguities of sustainability and at the same time offer a rational, scientific, and justifiable means to analyse and implement it. Froger and Munda (1998) call this framework procedural rationality, which is distinct from traditional scientific management with its strong commitment to quantification and rigid analysis.

In forestry, much of the MCA-related work has been led by the Center for International Forestry Research. In a series of articles, Mendoza and Prabhu (2000a-d) described different MCA models designed to measure C&I at different levels, and ultimately to evaluate forest sustainability at different scales; i.e. nationally, regionally, or at the forest management unit levels. These publications also reported on applications of MCA at different contexts under different organizational and institutional settings. Mendoza and Prabhu (2000a) described a sustainability assessment of a forest management unit that is commercially managed. On the other hand, Mendoza and Prabhu (2000b) examined the assessment of the sustainability of community managed forests.

Lessons learned from this series of case studies show that forest sustainability assessment, despite its inherent complexity, can be formalized and conducted systematically, because MCA has proven itself to be an effective tool for selecting, evaluating and aggregating the various indicators of forest sustainability. The studies show that MCA offers several characteristics that make it a suitable tool for C&I assessments: 1) it enhances the participatory approach to decision making, where all stakeholders are involved not only as information providers but also as decision makers; 2) it simultaneously accommodates different criteria; 3) it enables analysis involving mixed data, both qualitative and quantitative, including expert opinions in the absence of hard data; 4) it is highly transparent to all participants.

In addition to the research described above, a number of other authors and practitioners have adopted similar MCA-related studies and applications dealing with sustainability assessments under different contexts. Varma et al. (2000) described a methodology organized as a decision support system for sustainable forest management. The system integrates and utilizes spatial and temporal data that combine ecological, economic, and social indicators. A multi-criteria evaluation technique that uses a similar model described in the five case studies by Mendoza and Prabhu (2000a-d) was used to estimate the relative importance values of the different indicators. The model of Varma et al. (2000) directly incorporates uncertainty in the analysis, through the use of fuzzy approaches and the Dempster-Shafer theory of evidence. Directly accommodating uncertainty into the model enabled Varma et al. (2000) to structure sustainability analysis beyond determining relative weights of C&I. In addition, the model was used as a framework for participatory monitoring of sustainable forest management.

Ducey and Larson (1999) described a fuzzy approach in evaluating sustainability. Their work included many of the complex issues surrounding sustainability assessments such as lack of data, achieving balance between long-term and short-term goals, and diverse and possibly conflicting stakeholder goals. To demonstrate the applicability of their model, an illustrative example was presented describing how fuzzy multi-criteria alternatives can be evaluated.

Returning to the studies of Mendoza and Prabhu cited above, there are three MCA methodologies that were used in C&I assessment: 1) Pairwise comparisons, 2) Ranking, and 3) Rating. In a nutshell, the pairwise comparisons approach distils the complex C&I decision problem into a series of one-on-one comparisons on the significance of each pair of indicators relative to the relevant criterion in the hierarchical level above it. Ranking, on the other hand, is different from the pairwise comparison method in that C&I elements are not explicitly compared one-on-one. Instead, they are simply ranked according to their perceived importance. Finally, in rating, the indicators are judged by their relative degrees of importance indicated by "scores" instead of cardinal rank. Described briefly below are the three most common MCA methodologies, focusing particularly on the procedures for determining relative weights of each element.

Ranking method

This method involves analysis of decision elements, in which each element is assigned a "rank" depending on its perceived importance. Ranks are assigned following a 9-point scale as follows: 1 - *weakly important*; 3 - *moderately important*; 5 - *important*; 7 - *very important*; 9 - *extremely important*. Intermediate assessment values of 2, 4, 6 and 8 could also be used. The relative importance or weight can be calculated based on the ranks assigned to each element. For example, consider a criterion j with m indicators as described below:

$$C_j \in \{ I_{j1}, I_{j2}, \ldots, I_{jm} \} \qquad (1)$$

Suppose an expert k gave the following ranking to the respective indicators of criterion j as follows: $r_{jk1}, r_{jk2}, \ldots, r_{jkm}$ The relative weight, for indicator i; w_{ji} ($i = 1,2,\ldots, m$) can be calculated as follows:

$$w_{ji} = \sum_k r_{jki} / \sum_i \sum_k r_{jki} \qquad (2)$$

Rating method
This method directly assigns weights explicitly to each decision element by distributing 100 points, which is the sum of all weights. Hence, the weights of all decision elements must satisfy the following requirements:

$$0 \leq w_{ji} \leq 100; \text{ and } \Sigma \ w_{ji} = 100 \text{ for all } i \quad (3)$$

Pairwise comparisons
The pairwise comparison technique is the most involved and also provides the most information. Because the "core" of C&I assessment is the evaluation of the indicators, this method will be used to examine the relative weights at the indicator level. The technique is based on the method developed by Saaty (1995) called the Analytic Hierarchy Process (AHP). Saaty (1995), Mendoza (1997) and Mendoza and Prabhu (2000a) describe the procedural details of AHP along with a number of applications. This section briefly introduces the method only to highlight its applicability and suitability in the context of C&I assessments. For more details on the principles of AHP and its applications to natural resource and environmental management, readers are referred to Schmoldt et al. (2001).

AHP can be summarized as a 4-step procedure described as follows: Step 1, set up the decision hierarchy by decomposing the problem into a hierarchy of interrelated elements; Step 2, generate input data consisting of comparative judgment (i.e. pairwise comparisons) of decision elements; Step 3, synthesize the judgments and estimate the relative weights; and Step 4, determine the aggregate relative weights of the decision elements to arrive at a set of ratings for the decision alternatives. Procedures for estimating these relative weights are described in other published literature such as Saaty (1996) and Mendoza (1997).

In addition to estimating the relative weights, the AHP method also provides a measure of consistency called the consistency index. This index provides pertinent information about the consistency of the pairwise comparisons. Consistency is measured not only in terms of which element of the pair is more important but also how much more important. That is, consistency is reflected in both the ordinal and cardinal sense.

2.3. Scoring of indicators and a sustainability index measure

Perhaps the most significant, and also most difficult, part of sustainability assessment is to transform empirical data with respect to an indicator into an evaluative score or rank. This will require a consensus amongst those carrying out the assessment on how to scale the possible range of values measured empirically. This is inherently a subjective, value-laden step. However, it should be informed by current scientific knowledge as far as possible and must also reflect societal goals. Once agreement on how transformation of empirical data into scores has been reached it is recommended that a bottom-up approach be applied in that indicators are scored and evaluated before moving on to the next level up in the hierarchy, i.e., criteria and so on. This process of stepping up the hierarchy can be easily followed by referring to Figure 1. Again, it should be noted that sustainability has multiple dimensions; hence, the 'score' should be consistent with and meaningful for the dimension of sustainability for which the forest is being assessed.

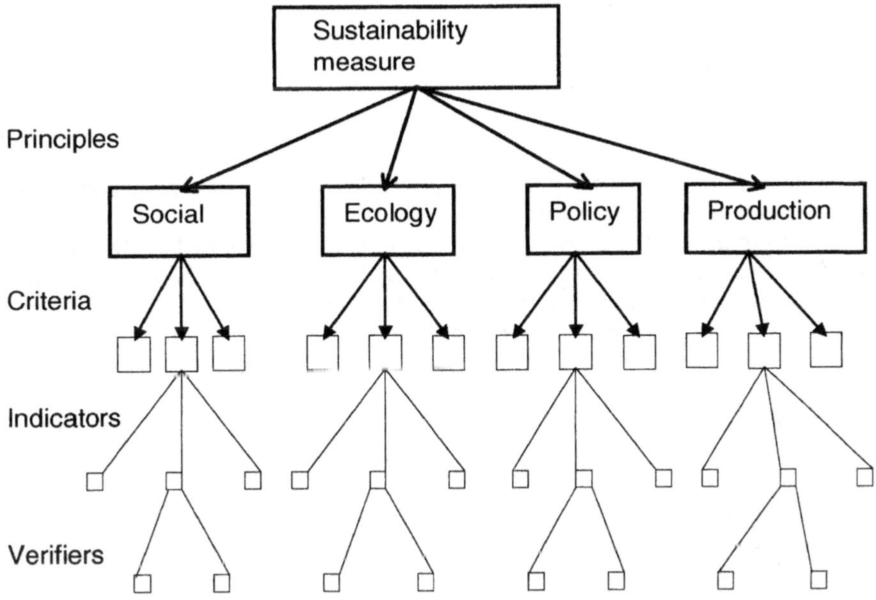

Figure 1. Hierarchical structure of C&I. This figure is not complete and is used only to lay out the components of the C&I hierarchy. Hence, the blank boxes are included to denote a set of C&I elements that are too many to include in one figure. The complete set of these principles, criteria, indicators and verifiers are contained in Prabhu et al. (1998).

Since the indicators are the measurable or tangible elements of this sustainability hierarchy, scoring should be done at this level. Scores or performance values of criteria and principles will be estimated based on the weights and scores of the indicators. That is, scores of higher-level concepts are aggregated from the lower level concepts.

In general, it may be instructive to estimate the overall condition or status of a forest as reflected by the sustainability measures or the indicators, criteria, and principle. This can be estimated by aggregating the sustainability measures (weights and scores) of indicators, criteria, and principle. The following model can be used:

At the criteria level:

$$SIC_i = \Sigma\, s_j w_j \tag{4}$$

where SIC_i is the sustainability index of criterion i, s_j is the score of indicator j and w_j is the relative weight of indicator j.

At the principle level:

$$SIP_p = \Sigma\, SIC_i w_i \tag{5}$$

where SIP_p is the sustainability index of principle p, SIC_i is the sustainability index for criterion i and w_i is the relative weight of criterion i.

Note from the models above that 'raw' scoring is only required at the indicator level. However, relative weights are estimated for both the indicators and the criteria. At the criteria level, 'scores' are simply aggregated from the indicators. Likewise, scores at the principle level are aggregated from the criterion scores.

3. PERSPECTIVES OF SUSTAINABILITY AND THRESHOLD VALUES

Sustainability is not only measured in terms of multiple criteria and indicators; it can also be viewed from different perspectives. One common view, for example, is the use of *critical thresholds*. Based on the value of the critical threshold of a particular indicator, and its current condition, inferences can be made about its sustainability status. If the current status of the indicator is close to its critical threshold, then the indicator is also considered critical and has low sustainability value. On the other hand, if the current value of an indicator is far from its critical threshold, then the indicator is far from being critical and will likely have high sustainability value.

Another view may be *sensitivity* of the indicator to change. Some indicators may be stable, resistant, or resilient. Hence, the indicator is likely to be able to absorb change and sustain itself. This indicator then is likely to have high sustainability value.

Finally, *risk* and *predictability* should also be considered in making inferences about sustainability. Indicators that are less predictable, and at the same time are risky in the sense that they could lead to adverse consequences if mismanaged, must be examined carefully.

3.1. Critical threshold values

Critical thresholds values (CTV) serve as limits within a resource utilization space. In this context, CTVs represent reference values associated to an indicator depicting allowable space for utilization. In an operational or practical sense, for as long as the value of an indicator falls within the bounds defined by the CTVs, the ecosystem is deemed sustainable. On the other hand, if the value of the indicator falls outside the CTV limits, this implies going beyond the flexibility of an ecosystem. The limits are called 'critical' because once the value of the indicator is exceeded, the ecosystem is judged as unsustainable with possible adverse consequences.

Conceptually, CTVs are easy to comprehend. Operationally however, they are extremely difficult to measure. With sufficient data and complete knowledge about ecosystem functions and processes, these limits could be specified with some certainty, but in practise they are seldom known.

Invariably, CTVs are difficult if not impossible to determine accurately. Specifying a CTV, particularly in a context as complex as a forest ecosystem, will always be fraught with uncertainty. Such uncertainty is not only statistical, e.g., prediction error, but also embraces the following concepts: 1) *generality* where uncertainty derives from differences due to the wide variety of situations, 2) *ambiguity* where uncertainty embraces more than one distinguishable sub-concept, and 3) *vagueness* because precise boundaries are not well defined.

In view of the extensive amount of uncertainty described above, it is more meaningful to specify a CTV, not as a point estimate, but as a range of values expressed as CTV_{min} and CTV_{max}. One can view the range as mirroring the maximum and minimum CTVs expressed by experts where CTV_{min} indicates a conservative estimate of the allowable threshold, while CTV_{max} reflects a more liberal (or risky) estimate of the allowable threshold.

3.2. Qualitative assessment of critical thresholds

Nijkamp and Vreeker (1999) described a simple methodology for analysing CTVs in the context of assessing sustainability. The method has qualitative features that lend themselves well to problems that are characterized by high degrees of ambiguity, uncertainty, subjectivity and general lack of information. The method is referred to as Flag Theory (Nijkamp and Ouwersloot, 1997) because the assessment process involves assigning colored flags to the indicators depending on their critical thresholds. For example, consider one indicator whose critical values can be described in a continuum as shown in Figure 2. Nijkamp and Vreeker (1999) proposed that the line segment of Figure 2 could be interpreted as:

- Area A: Green Flag: No reason for specific concern
- Area B: Orange Flag: Be alert

- Area C: Red Flag: Reverse trend
- Area D: Black Flag: Stop immediately. Remedy urgently needed.

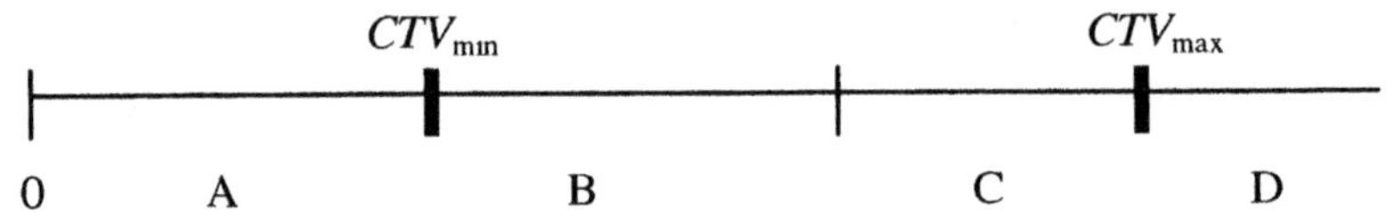

Figure 2. An example of splitting the range of variation of an indicator into different flag areas (A, B, C and D)

The flagging approach described above offers an intuitive way to make the assessment –a procedure that is highly transparent to experts. Operationally, each indicator will be evaluated by assigning colored flags according to their perceived degree of criticality as shown in Figure 2.

3.3 Quantitative assessment of critical indicators

The approach described above is a qualitative way of assessing sustainability using coloured flags in lieu of quantitative values. Another approach is to treat uncertainty surrounding the threshold values as a fuzzy concept. In other words, threshold values are treated as fuzzy critical thresholds and can be analysed using the theory of fuzzy sets.

Zadeh (1965) introduced fuzzy sets as a means for dealing with uncertainty. Uncertainty in the context of fuzzy sets implies inexactness or imprecision, not in terms of randomness or probability. In other words fuzzy sets, in contrast with classical set theory, can be applied to a different kind of uncertainty where elements, for various reasons, can not always be defined sharply because of generality, ambiguity, and vagueness, as described in Section 3.1.

Intuitively, a fuzzy set is a defined collection of elements which can be represented by natural (fuzzy variable) or numeric (fuzzy value) language. The definition of a set can be represented using the membership function $\mu_A(x)$. The membership function of a fuzzy set defines the grade of membership of x belonging to the fuzzy set A. Usually, membership function values range from 0 to 1, 1 representing full membership in the set and 0 non-membership. The grades of membership of x in A reflect partial ordering that is not based on probability but on admitted possibility. The membership function of a fuzzy set defines how the grade of membership of x in A is determined. It is used to describe the 'vague', general, or ambiguous characteristic of the fuzzy set. In the context of the sustainability assessment, the membership function essentially reflects the degree of sustainability of an indicator or criterion. Hence, sustainability index (SI) in terms of the membership function can be formulated as:

$$\mu_A(x) = \begin{cases} 0, \text{if } x \leq CTV_{min} \\ [CTV_{max} - x]/[CTV_{max} - CTV_{min}] \text{ if } CTV_{min} < x < CTV_{max} \\ 1, \text{if } x > CTV_{min} \end{cases} \tag{6}$$

where x is the current value of indicator.

Given the membership function described in (6), the sustainability index model described in (4) and (5) can be modified as:

$$\text{SIC} = \Sigma\, \mu_j w_j \tag{7}$$

where μ_j is the membership function of indicator j.

3.4. Compensatory and non-compensatory thresholds

The model in (4) and (5) assume that the indicators are compensatory; that is there can be trade-off among the indicators. Hence, a low SI value for one indicator may be 'compensated' by other indicators so that at the aggregate criterion level, the SI may be considered sustainable despite the low SI value, possibly even critical, of one or more indicators. In other words, it is possible to have a criterion SI value that is sustainable although at least one indicator is critical (i.e., its value is below its CTV_{min}). This may be acceptable if the critical threshold is of a lower-weighted indicator. On the other hand, this situation may not be acceptable if the critical thresholds include very significant or highly weighted indicators. In this case, the 'thresholds' of the significant indicators determine the 'binding' threshold for estimating the SI value at the criterion level. Hence, if one or more indicators are non-compensatory (e.g., highly weighted or primary indicators), then the sustainability index model in (4) is modified as follows:

$$\text{SIC} = \sum s_j w_j \prod c_i \tag{8}$$

where c_i are the 'binding' indicators (i.e. important indicators). It is equal to 1 if the threshold value of indicator i is satisfied, and 0 if the threshold value is not satisfied.

The model described in (8) is non-compensatory in the sense that SI drastically takes a value 0 at the threshold point of any one of the primary indicators i. At any other point, the model is compensatory in the sense that other indicators can compensate (i.e., trade-off is possible) for each other so that potentially low SI of one indicator j may be improved by high SI values of other indicators. In some situation where primary indicators serve as essentially dominant indicators or critical indicators and other indicators are considered secondary, the model in (8) may be most appropriate. Hence, sustainability analysis is driven primarily by the conditions of the primary indicators.

4. ASSESSMENT OF CROSS-INDICATOR INTERACTIONS

4.1. Indicator linkages

Criteria and indicators are often interlinked and, consequently, may impact sustainability directly or indirectly. It is conceivable that although assessment of individual indicators appears to be good, forest sustainability may be compromised due to cross-criterion or cross-indicator interactions. These interactions may not be apparent if criteria or indicators are assessed without examining their interrelationships between and among each other as Prabhu et al. (2001) have pointed out.

Besides assessing sustainability through the interactions or linkages among indicators, it is also conceivable that ecosystem risk is affected not only by the risk associated to each indicator, but also the ecosystem-wide risk that emanates or emerges from cross-indicator linkages and interactions. Hence, while an indicator in and by itself may not be at risk, it may eventually be compromised because of its linkages either directly or indirectly with other indicators that are at risk.

One systematic way to analyse cross-indicator linkages is through the use of a mathematical sub-discipline called graph theory. As the name implies, graphs are used to denote the relationships using nodes and arrows, where the direction of the arrow represent the direction of the impact or the nature of the causal relationship. In addition, either positive or negative feedback loops can be used to characterize the nature and direction of causalities among indicators, using carefully designed arrow diagrams between and among indicators.

4.2. Extent of connectivity: domain and potency of indicators

One of the simplest measures of the extent of connectivity or linkage of an indicator is the number of indicators directly linked to it. This may be referred to as the 'domain' of an indicator; that is, the density of indicators directly linking in and out of the indicator. Indicators can then be ranked according to their 'potency', i.e. the number of linkages.

Another useful concept is called the 'critical domain'; i.e. the number of 'critical indicators' that are linked to an indicator. Closely related to 'critical domain' is the concept of 'critical path' – a path consisting of critical indicators linked directly to each other. Perhaps one of the most significant aspects of ecosystem risk analysis is the number of critical indicators (i.e. those indicators that are at risk) directly linked to one another. This is important because an indicator that may currently be in good condition can potentially be placed at risk because of its connection with, and the possible negative impacts of, critical indicators. 'Criticality' can be viewed in terms of the perceived risk status associated with each indicator.

In addition to potency analysis (domain, critical domain, or critical path), another insightful analytical construct is the concept of 'centrality' of an indicator. Central analysis goes further than domain analysis in that it calculates 'central values' to reflect not only those indicators that are immediately linked to the specific indicator but also those which link through them. Centrality scores will reflect not only the number of direct and indirect linkages, but also the level (i.e. how far removed the indirect connection is).

4.3. Graph-theoretic analysis of cross-indicator interactions

Cognitive mapping techniques and qualitative influence diagram methods are two alternative models that can be used to analyse cross-indicator linkages and interactions. For many forest sustainability assessment contexts, especially those dominated by qualitative criteria and indicators (e.g. socio-economic and policy-related C&I), these approaches are perhaps the most appropriate analysis tools. In some situations, it may be desirable to move the analysis beyond the determination of potency or domain, critical domain and paths, and centrality of the indicators.

More formal analysis of arrow or influence diagrams is possible through graph theory. Wenger et al. (1999) reported one of the pioneering studies on graph-theoretic analysis of relationships among ecosystem stressors. The study recognized the need to develop analytical tools for holistic assessment of ecosystem risk due to stressors that threaten an ecosystem's viability or integrity. This study is significant because it is the first to use graph theory to analyse the interactive relationships among stressors. Graph theory allows more structured analysis of the linkages, the manner in which stressors are connected, and the interactive paths among stressors (i.e. the feedback loops), which can augment or diminish the impacts of some stressors. Results of graph-theoretic analysis can help identify strategically significant clusters of stressors that should be closely monitored.

The parallel between the graph-theoretic analysis of stressors to assess ecosystem risk and the assessment of forest sustainability is quite intuitive. Both are organized and structured using directed graphs. The criteria and indicators are analogous to stressors; the linkages and connectedness are represented by the directed graphs.

A detailed description of the graph theory approach is contained in Wenger et al (1999), and will not be rehashed here. However it is worth noting that some of the insights and concepts that can be drawn from this pioneering study are meaningful to sustainability analysis. One such concept is the adjacency matrix $\mathbf{A} = (a_{ij})$ defined as follows (S stands for 'stressor'):

$$a_{ij} = \begin{cases} 1, \text{if } S_i \text{ has an impact on } S_j \\ 0, \text{if } S_i \text{ has no impact on } S_j \end{cases} \tag{9}$$

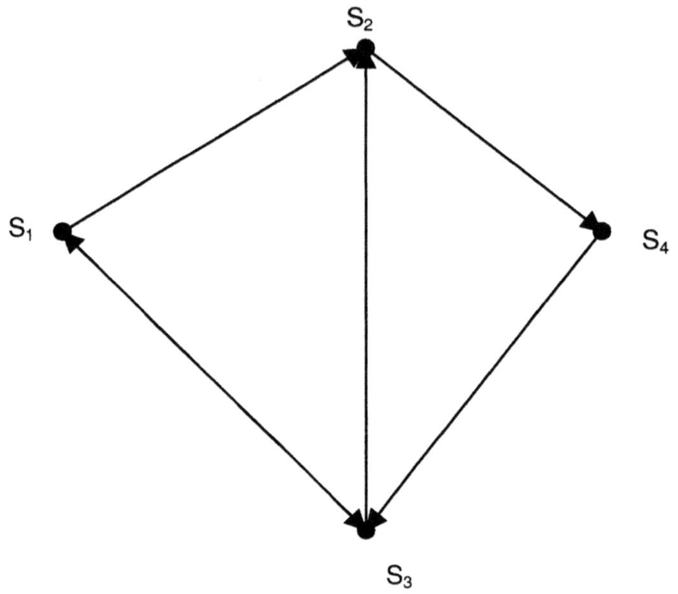

Figure 3. A hypothetical digraph of ecosystem stressors. The arrows indicate the directions of impacts. The graph is replicated from Wenger et al. (1999).

To illustrate the adjacency matrix and the other meaningful graph-theoretic concepts, the sample arrow diagram described in Wenger et al. (1999) is replicated in Figure 3. Based on this simple 4-node arrow diagram, the adjacency matrix is as follows:

$$\mathbf{A} = \begin{vmatrix} 0 & 1 & 1 & 0 \\ 0 & 0 & 0 & 1 \\ 1 & 1 & 0 & 0 \\ 0 & 0 & 1 & 0 \end{vmatrix} \tag{10}$$

Closely associated with the adjacency matrix **A** is another useful concept called reachability matrix **R**; another hierarchy matrix with values 1 (if S_j is reachable from S_i) or 0 if S_j is not reachable from S_i). Using Figure 3, the reachability matrix **R** is as follows:

$$\mathbf{R} = \begin{vmatrix} 1 & 1 & 1 & 1 \\ 1 & 1 & 1 & 1 \\ 1 & 1 & 1 & 1 \\ 1 & 1 & 1 & 1 \end{vmatrix} \qquad (11)$$

Hence, based on the matrix above, all nodes can be reached from any node. This can easily be verified by examining the connections in Figure 3. The matrix also implies that every stressor impacts every other one directly or indirectly.

Wenger et al. (1999) have also shown that the reachability matrix can be determined computationally using the formula:

$$\mathbf{R} = \mathrm{B}[(\mathbf{I} + \mathbf{A})^{n-1}] \qquad (12)$$

where $\mathbf{I}$ is the $n \times n$ identity matrix and B is a Boolean function. The model in (12) allows the estimation of the reachability matrix for more complex sets of digraphs since it can be computed from the adjacency matrix $\mathbf{A}$.

Based on these two constructs, it is also possible to establish the strength of connectivity of the digraphs. From strongly connected digraphs, clusters of strong components can be identified; that is, subsets of stressors or indicators which are strongly linked together. These clusters are significant because they are metaphorically the 'backbone' of a set of digraphs, which in turn represents the set of stressors or indicators considered significant in a forest ecosystem. These clusters then should be the focus of monitoring efforts because they are pivotal in the functioning of an ecosystem. Their sphere of influence is more extensive compared to the others.

Here and in the previous section we have briefly described how cognitive mapping using influence diagrams and graph theory using digraphs can be used complimentarily to closely examine cross-indicator linkages and interactions. Insights gained from these two analytical procedures can be helpful towards better understanding of the forest ecosystem in particular, and forest sustainability in general.

5. SAMPLE APPLICATIONS

The models described in the previous sections lend themselves well to sustainability assessments under different contexts and objectives. Mendoza and Prabhu (2000a-d, 2002) describe case studies with different sustainability scenarios. Because of space limitations, it is not possible to provide application scenarios that illustrate all the models described in this paper. Nonetheless, three illustrative examples are described in the succeeding sections to highlight the features of some of the models described earlier.

5.1 Sustainability index and critical thresholds

To illustrate the application of the models described in (1) and (4), part of the case study reported by Mendoza et al. (2001) is briefly described in this section. One of the criteria of forest sustainability as reported in the previous study is stated as: "*Ecosystem function is maintained*". Under this general criterion, four indicators were specified as shown in Table 1. In this case study involving a community-managed forest, an assessment team consisting of different stakeholders was constituted and served as the evaluators that provided sustainability scores as described in (4). All members of the assessment team voted individually; their votes were aggregated as shown in Table 1. In addition, the ranking method described in (1) was used to estimate the relative weights shown in the table. These estimates of sustainability scores and relative weights of all four indicators are used to calculate the overall sustainability index of the criterion.

Given the scoring scale used (i.e., 1 if the indicator is close to its critical value and therefore its prospect for sustainability is low; 5 if the indicator is far from its critical value and therefore its prospect for sustainability is high), all the indicators were judged as performing well with their average scores better than 3.5. The criterion was judged as moderately sustainable, with a criterion score of 3.64, calculated using the relative weights and sustainability scores from Table 1. To illustrate the use of the model in (4), the aggregated criterion score was calculated as: $0.29 \times 3.67 + 0.24 \times 3.67 + 0.29 \times 3.5 + 0.17 \times 3.8 = 3.64$. Note that the fourth indicator is deemed less important relative to the other three. Hence, the contribution of this indicator to the overall sustainability of the criterion is relatively less compared to the others.

Table 1. Relative weights and sustainability scores of indicators of the sustainability criterion 'Ecosystem function is maintained'.

Indicator	*Relative weight*	*Sustainability score*
Ecologically sensitive areas, especially buffer zones and riparian strips along watercourses, are protected	0.29	3.67
Representative areas, such as critical watershed, waterfalls, caves, and ecotourism areas and other sites of ecological importance are protected and appropriately managed	0.24	3.67
Erosion and other forms of soil degradation are minimized	0.29	3.50
No chemical contamination is introduced into food chains	0.17	3.80

5.2. Fuzzy approach to critical threshold value assessments

To demonstrate the application of the fuzzy model described in (7), the case study reported by Mendoza and Prabhu (2001) is used as an illustrative example. The example that follows constitutes only a part of the case described in the previous study which consists of assessing the Biodiversity Conservation Index as one of the criteria for forest sustainability. Under this criterion, there are four indicators, and eight verifiers under these four indicators. The indicator and verifier hierarchy is shown in Figure 4. Again, an expert team was constituted and served as the primary sources of information and evaluation data. AHP was used to estimate the relative weights of each verifier relative to an indicator (e.g. indicator 2 has more than 2 verifiers so the pairwise comparison is appropriate), and the indicators relative to the biodiversity criterion.

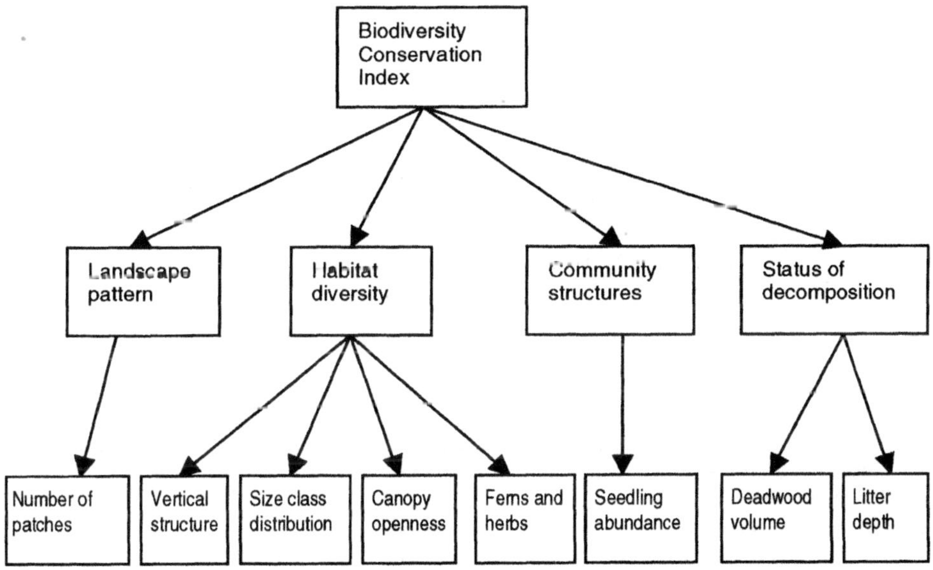

Figure 4. Criteria and indicator hierarchy for biodiversity conservation.

The data used in this example were obtained from 12 plots located in a 3,000-hectare forest in Indonesia. The summary statistics, particularly the mean of each fuzzy verifier, are shown in Table 2. Table 3 shows the relative weights, based on the use of AHP, for both the verifiers and the indicators. CTV limits of each fuzzy verifier, as described in (6) are also shown in Table 3. Membership function values for the verifiers, calculated using the weights and CTV limits, are shown in Table 4. Finally, using the estimated membership function values of each verifier, and the weights associated with the verifiers and indicators, in models (6) and (7), the Biodiversity Conservation Index for each plot is calculated, and summarized as in Table 5.

Table 2. Means of fuzzy verifier data obtained from 12 sample plots.

Plot	*No. of patches*	*Vertical structure (m)*	*Size classes (cm)*	*Canopy openness (%)*	*Herbs*	*Seedlings per ha*	*Depth of litter (cm)*	*Dead-wood volume (m^3ha^{-1})*
1	2	21.02	23.29	78.83	2	380	3.2	19.21
2	1	24.66	21.50	84.58	3	560	3.4	24.21
3	3	22.69	21.49	87.95	5	601	4.1	23.73
4	2	20.19	23.43	82.21	3	568	2.2	17.07
5	1	18.97	21.68	73.71	4	467	4.6	35.42
6	1	21.11	22.81	72.75	6	447	5.4	22.43
7	2	20.86	23.89	79.67	3	567	5.6	33.59
8	2	16.69	17.46	83.54	4	678	2.5	18.82
9	2	22.11	24.99	80.29	6	326	3.6	16.90
10	1	17.33	22.11	89.17	4	435	4.3	31.46
11	2	18.53	54.38	85.13	4	478	3.8	35.32
12	2	14.67	16.02	80.79	6	562	3.8	55.07

Source: Mendoza and Prabhu (2001)

Table 3. Relative weights and membership function limits of verifiers and indicators.

Indicator / Verifier	*Relative weight*	CTV_{min}	CTV_{max}
1: Landscape pattern	**0.28**		
Number of patches	1.00	0	4
2: Habitat diversity	**0.28**		
Vertical structure	0.41	5	20
Size class	0.26	5	20
Canopy openness	0.25	50	80
Herb community*	0.09	1	10
3: Community structures	**0.24**		
Seedling abundance	1.00	360	600
4: Status of decomposition	**0.21**		
Deadwood volume	0.47	10	40
Depth of litter	0.53	0	10

****Herb community is classified as a categorical data between 0 and 10; each category reflects the amount of herbs (e.g., scarce, scattered, 5% of plant cover, 20% of plant cover, etc.)***

Table 4. Membership function values (μ) of each plot based on mean values of verifiers.

	Number of patches		Vertical structure (m)		Size classes (cm)		Canopy openness (%)		Herbs		Abundance of seedlings		Depth of litter (cm)		Deadwood volume (m^3ha^{-1})	
Plot	Mean	μ	Mean	μ	Mean	μ	Mean	μ	Mean	μ	Mean	μ	Mean	μ	Mean	μ
1	2	0.50	21.02	1.00	23.29	1.00	78.83	0.96	2	0.11	380	0.60	3.2	0.32	19.21	0.30
2	1	0.25	24.66	1.00	21.50	1.00	84.58	1.00	3	0.22	560	0.92	3.4	0.34	24.21	0.47
3	3	0.75	22.69	1.00	21.49	1.00	87.95	1.00	5	0.44	601	1.00	4.1	0.41	23.73	0.45
4	2	0.50	20.19	1.00	23.43	1.00	82.21	1.00	3	0.22	568	0.94	2.2	0.22	17.07	0.23
5	1	0.75	18.97	0.93	21.68	1.00	73.71	0.79	4	0.33	467	0.76	4.6	0.46	35.42	0.84
6	1	0.50	21.11	1.00	22.81	1.00	72.75	0.75	6	0.55	447	0.72	5.4	0.54	22.43	0.41
7	2	0.25	20.86	1.00	23.89	1.00	79.67	0.98	3	0.22	567	0.94	5.6	0.56	33.59	0.78
8	2	0.75	16.69	0.77	17.46	0.83	83.54	1.00	4	0.33	678	1.13	2.5	0.25	18.82	0.29
9	2	0.50	22.11	1.00	24.99	1.00	80.29	1.00	6	0.55	326	0.51	3.6	0.36	16.90	0.23
10	1	0.75	17.33	0.82	22.11	1.00	89.17	1.00	4	0.33	435	0.70	4.3	0.43	31.46	0.71
11	2	0.25	18.53	0.90	54.38	1.00	85.13	1.00	4	0.33	478	0.78	3.8	0.38	35.32	0.84
12	2	0.75	14.67	0.64	16.02	0.73	80.79	1.00	6	0.55	562	0.93	3.8	0.38	55.07	1.00

Source: Mendoza and Prabhu (2001)

Table 5. Indicator sustainability values and the biodiversity conservation index of plots.

Plot	Landscape pattern	Habitat diversity	Community structures	Status of decomposition	Biodiversity Conservation Index
1	0.50	0.92	0.60	0.32	0.6096
2	0.25	0.94	0.92	0.37	0.6317
3	0.75	0.96	1.00	0.42	0.8070
4	0.50	0.94	0.94	0.22	0.6750
5	0.75	0.86	0.76	0.54	0.7466
6	0.50	0.91	0.72	0.51	0.6747
7	0.25	0.94	0.94	0.61	0.6869
8	0.75	0.81	1.13	0.26	0.7626
9	0.50	0.97	0.51	0.33	0.6033
10	0.75	0.87	0.70	0.49	0.7245
11	0.25	0.91	0.78	0.48	0.6128
12	0.75	0.75	0.93	0.67	0.7839

calculated using (6) as follows: $\mu_x = [4\text{-}2]/4 - 0$, where $CVT_{max} = 4$ and $CTV_{min} = 0$ as shown in Table 3. The mean value of the verifier is equal to 2 as shown in Table 2.

The membership functions of the other verifiers were calculated in the same manner as shown in Table 4. Given the membership function values of each verifier, the SI values of the indicators are aggregated based on the model in (7). For example, consider the indicator 'habitat diversity' that has four verifiers. Given the weights of these verifiers shown in Table 3, and their respective membership functions in Table 4, the SI value of this indicator is calculated as follows: SI = 0.41 × 1 + 0.26 × 1 + 0.25 × 0.96 + 0.09 × 0.11 = 0.92. The SI values of the other indicators are estimated following the same procedure. Finally, the Biodiversity Conservation Index of Plot 1 is estimated by aggregating the SI values of the four indicators. This is done using the model in (5) given the relative weights of the indicators in Table 3 and their respective SI values in Table 5 as follows: BSI = 0.28 × 0.5 + 0.28 × 0.92 + 0.24 × 0.60 + 0.21 × 0.32 = 0.609.

The biodiversity conservation index values contained in Table 5 are, fundamentally, based on the assessments of the fuzzy verifier sustainability. Note that the index values reflect the degree of sustainability; hence, the higher the value of the index, and the closer it is to the value 1, the more sustainable the forest is judged to be. The opposite is also true; that is, a lower index value implies a less sustainable forest.

5.3. Cross-indicator assessment

Section 4 presented an expository overview of two procedures that can be used to examine cross-indicator interactions. Analysing the interactions of the indicators can be quite insightful because it gives a broader picture of the overall impact of all indicators. To illustrate this analysis, the study described in Mendoza and Prabhu (2002) is presented as a demonstrative example.

The example involves a community-managed forest located in Zimbabwe. Management of the forest is vested on a Resource Management Committee selected locally. As is the case in the two other examples described above, an expert team comprised mainly of locally based experts was constituted to serve as an assessment team to evaluate the sustainability of the community forest. The team was first asked to develop a set of criteria and indicators. The team then examined the generated set of indicators and mapped out their causality relationships (if known) using directed graph (or graph) diagram as described in Section 4. The team generated 51 indicators covering a variety of concerns including social, economic, biodiversity, ecology, and policy issues. Not all of the 51 indicators were sufficiently known, understood, or clear to be included in an interaction analysis. Consequently, only 30 indicators were included in the cross-indicator linkage analysis as shown in Table 6.

Table 6. Cross-indicator interaction analysis.

Tail of arrow (causal indicator)	*Head of arrow (affected indicator)*	
	Within criteria	*Outside criteria*
To improve and maintain the Mafungautsi biodiversity		
1.1 Number of animals and plants against carrying capacity	1.2	
1.2 No of locally endangered species		
1.3 No of added/new species (trees, animals plants etc) *		
1.4 Incidence of veld fires	1.1/1.5	2.1/2.2/2.3
1.5. Variety of plant and animal species	1.1	5.6/5.10
To Enhance the ecological and biological function of Mafungautsi		
2.1 Forested area	2.2	4.8
2.2 Tree canopy cover		4.8
2.3 Availability of aroma, dustless atmosphere, fresh wind		
2.4 Amount of fragmentation (clearing)	2.1	
2.5 Rainfall pattern	2.1	4.3/4.5
2.6 Rainfall intensity	2.7	4.3/4.7
2.7 Amount of soil washed away		3.12
2.8 Annual temperature variations		
To enhance livelihood of communities and the nation		
3.1 Diversity of income sources and levels	3.8/3.7	
3.2 Family size		
3.3 Sources of energy		2.1
3.4 Improved road infrastructure and communication		
3.5 Improved level of literacy within the community		
3.6 Improved quality of accommodation for households		
3.7 Assets owned		
3.8 Number of cattle	3.8	1.4/2.5
3.9 Household annual food reserves in relation to family size		
3.10 Less infant mortality		
3.11 Availability of safe water and sanitation		
3.12 Occurrence of prohibited human activities		1.4/2.1
3.13 Open and equitable participation in decision and rules making		

Table 6 (Continuation). Cross-indicator interaction analysis.

Tail of arrow (causal indicator)	*Head of arrow (affected indicator)*	
	Within criteria	*Outside criteria*
To conserve the quality of the watershed		
4.1 High land productivity per unit area crop	4.2	1.1/1.5/3.7/3.1
4.2 Level of the water table	4.7	
4.3 Flowing water		
4.4 Decomposition rate/presence of dung beetles		
4.5 Existence of perennial pools		
4.6 Clean water		
4.7 Number of gullies and rills	4.1	
4.8 Litter cover good (micro climate)		
To Manage and utilize forest produce wisely		
5.1 Harvesting rate of timber		
5.2 Controlled livestock units		
5.3 Existence of rules to guide collection of resources	5.1/5.4/5.5/5.8/5.9	1.1/1.5/1.4
5.4 Harvesting area of broom grass		
5.5 Harvesting area of thatch grass		
5.6 Availability of herbs		
5.7 Gender sensitive activities		
5.8 Firewood quota		3.3
5.9 Amount of green wood coming from the forest		1.1/1.5
5.10 Availability of wood and non wood products		
5.11 Improved pricing of products	5.1/5.4/5.5/5.8/5.9	
To enhance socio-cultural values		
6.1 Exploitation of medicinal plants by local communities		
6.2 Improved/increased resource sharing activities		
6.3 Knowledge gap between elders and children on local species		
6.4 No of visitors to recreational centres		
6.5 Number of sacred sites		
6.6 Number of rituals performed at sacred sites		

**Shaded indicators are considered 'inactive' for the purpose of the cross-indicator interaction analysis. Inactive indicators have no arrows originating from, or ending at them. This does not imply they are not active or important. They are treated as inactive because: 1) there is insufficient knowledge about their linkages or relationships with other indicators; 2) they are not clearly defined, so that linkages with other indicators cannot be drawn; 3) they are vaguely stated so that their linkages with other indicators are not meaningful (e.g., socio-economic indicators); or 4) they are not compatible or properly matched (e.g., social indicators with biological indicators)*

Based on this linkage matrix, some of the measures of influence and overall interaction effects, such as domain and centrality measures, are evaluated. Some of these measures are contained in Table 7. The central score can be viewed as the strategic value of an indicator because it reflects not only the number of indicators impacted directly but also its overall reach in terms of its indirect connections with other indicators. The central score considers: 1) number of indicators directly linked to the indicator, and 2) number of indicators indirectly connected at different levels or nodes beyond the indicator's direct connections. Level 1 connections are direct; Level 2 is one connection removed; Level 3 is two connections removed, and so on. The impact of an indicator at higher levels (i.e. further removed from the direct connection) is necessarily lower. Hence, the central score of an indicator is adjusted as follows: 1) the number of indicators at Level 1 (S_1) is divided by 1, 2) the number of indicators at Level 2 (S_2) is divided by 2, and so on. Hence, the central score (C_i) of indicator i considering j levels of interactions can be calculated as follows (Mendoza et al, 2001b):

$$C_i = S_j/j + \ldots + S_n/n \; ; \quad j = 1, 2, \ldots n \qquad (13)$$

In terms of *Domain*, 7 of the 30 indicators (i.e., 1.1, 1.4, 1.5, 2.1, 4.1, 5.3, and 5.11) have a density of at least 5 indicators directly linked to them. These indicators are 'tactically significant' because of their potency, or the number of indicators they are connected with (i.e., affecting or affected by them).

The central scores in Table 7 were estimated using model (13) above based only on three levels (the default for the software used called Decision Explorer™ (Eden and Ackermann 1998). Of these, indicators 1.4, 1.5, 2.1, 4.1, 5.9, 3.12, and 3.8 have central scores of at least 10. As pointed out above, the central scores are indicative of the overall strategic importance of an indicator. Hence, these seven indicators can be perceived as the most strategic of the thirty whose impacts reach beyond their direct connections. Indicator 1.4 has the highest central score (15) and is therefore the most strategic. Indicator 1.5 has the next highest central score (14). Note that indicator 1.5 has more indicators linked to it. However, indicator 1.4 has a higher central score because it has more direct connections. While it may not be obvious in (13), the central score, and implicitly the strategic significance, is 'adjusted' as the level of indirect connection rises.

Table 7. Domain and centrality values.

	Domain (no. of indicators connected directly)	*Central scores*	*No. of indicators linked directly and indirectly*
To improve and maintain biodiversity in MSF			
1.1 Number of animals and plants against carrying capacity	5	14	27
1.2 No of locally endangered species	1	7	17
1.4 Incidence of veld fires	7	15	26
1.5 Variety of plant and animal species	5	14	27
To enhance the ecological and biological functions of MSF			
2.1 Forested area	7	13	23
2.2 Tree canopy cover	3	10	20
2.3 Availability of aroma, dustless atmosphere, fresh wind	1	7	17
2.4 Amount of fragmentation (clearing)	1	6	15
2.5 Rainfall pattern	4	10	20
2.6 Rainfall intensity	3	7	15
2.7 Amount of soil washed away	2	7	17
To enhance the livelihood of communities within MSF			
3.1 Diversity of income sources and levels	3	9	19
3.3 Sources of energy	2	9	21
3.7 Assets owned	2	6	13
3.8 Number of cattle	3	11	23
3.12 Occurrence of prohibited human activities	3	11	22
To conserve the quality of the watershed within MSF			
4.1 High land productivity per unit area crop	6	12	21
4.2 Level of the water table	2	6	14
4.3 Flowing water	2	7	16
4.5 Existence of perennial pools	1	4	12
4.7 Number of gullies and rills	3	8	16
4.8 Litter cover good (micro climate)	2	6	15
To manage and utilize forest produce wisely			
5.1 Harvesting rate of timber	2	5	9
5.3 Existence of rules to guide collection of resources	5	8	14
5.4 Harvesting area of broom grass	2	5	9
5.5 Harvesting area of thatch grass	2	5	9
5.6 Availability of herbs	2	7	9
5.8 Firewood quota	3	7	17
5.9 Amount of green wood coming from the forest	4	11	16
5.11 Improved pricing of products	5	8	22

Source: Mendoza and Prabhu (2002)

6. SUMMARY

This chapter described, in broad terms, different measures and approaches to assessing forest sustainability. In assessment contexts where data, information, and knowledge are significant, a formalized and structured model called MCA can be used to generate explicit measures or index values of sustainability. On the other hand, if assessments are clouded with uncertainty, or if data, information, or knowledge about the forest being assessed are indeterminate or even lacking, more qualitative approaches such as the use of flags can be used to broadly characterize the status or condition of the forest.

Fuzzy methods are another approach to handling uncertainty and the inherent complexity of forest sustainability assessments. The methods offer the flexibility to accommodate both qualitative and quantitative aspects of the assessment, as well as the lack of data or knowledge about the forest ecosystem itself. The fuzzy method is particularly appropriate in dealing with critical threshold values of indicators, which are generally unknown or difficult to determine. The ability to measure these limits as fuzzy thresholds allows the assessment process to be carried out using expert judgments on sustainability scores of criteria and indicators. Combining the concepts of critical thresholds and fuzzy set theory enables sustainability evaluation in terms of degrees of sustainability instead of a dichotomous assessment of whether forests are sustainable or not.

Indicators seldom impact forest sustainability by themselves independent of other indicators. Forest sustainability cannot be explained based on indicators alone, and while aggregating indicators through MCA may provide an overarching assessment of sustainability, this is hardly sufficient to reflect overall indicator impacts. In order to explain sustainability we need to take into account the dynamic interactions between indicators as well as their individual impacts. Consequently, indicators may affect overall sustainability indirectly through their impacts on other indicators. Again, because some indicators are quantifiable, they can be assessed using quantitative systems dynamics. This procedure allows more in-depth analysis of indicators including explicit critical threshold values. Other indicators are inherently qualitative and can only be evaluated using less structured procedures like graph theory or cognitive mapping techniques. These less structured models are designed to look more deeply into the dynamic interactions of indicators without requiring quantitative measures or explicit functional relationships among indicators. Insights gained from these qualitative systems dynamics procedures include assessments of the following concepts: a) indicators' 'domain' of influence, b) the strength of connectivity ('reachability') among and between indicators, and c) the 'centrality' of indicators. These insights can help identify indicators that must be protected and monitored closely because of their tactical and strategic influence in the overall functioning and dynamics of an ecosystem.

REFERENCES

AF&PA (American Forest & Paper Association). 1995. *Sustainable forestry: principles and implementation guidelines*. Washington D.C.

Bell, S. and Morse, S. 1999. *Sustainability Indicators: Measuring the Immeasurable*. Earthscan Publications Ltd. London.

Bockstaller, C., Girardin, P. and van der Werf, H.M.G. 1997. Use of agroecological indicators for the evaluation of farming systems. *European Journal of Agronomy* 7: 261-270.

Brlacich, M., Bryant, C.R and Smith, B. 1991. Review and appraisal of concept of food production systems. *Environmental Management*. 15(1): 1-14.

Committee of Scientists. 1999. *Sustaining the people's lands: Recommendations for stewardship of the national forests and grasslands into the next century*. Washington, D.C. US Department of Agriculture, Forest Service.

Conway, G.R. 1994. Sustainability in agricultural development: trade-offs with productivity, stability and equitability. *Journal of Farming Systems Research Extensions* 4(2): 1-14.

Cornforth, I.S. 1999. Selecting indicators for assessing sustainable land management. *Journal of Environmental Management* 56: 173-179.

Costanza, R., Norton, B. and Haskell, B.D. 1992. *Ecosystem Health: New goals for Environmental Management*. Island Press.

Cramb, R.A. 1993. Shifting cultivation and sustainable agriculture in East Malaysia: A longitudinal study. *Agricultural Systems*. 42: 209-266.

Ducey, M.J. and Larson, B.C. 1999. A fuzzy approach to the problem of sustainability. *Forest Ecology and Management* 115:29-40.

Eden, C. and Akermann, F. 1998. *Making Strategy: The journey of strategic management*. Sage Publications. London.

FSC (Forestry Stewardship Council). 1994. *Forestry Stewardship Principles and Criteria for Natural Forest Management*. Oaxaca, Mexico.

Froger, G. and Munda, G. 1998. Methodology for environmental decision support. In: Faucheux, S. and O'Connor, M. (Eds.). *Valuation for Sustainable Development*. Edward Elgar, Cheltenham, UK.

Gray, R. 1991. Economic measures of sustainability. *Canadian Journal of Agricultural Economics*. 39: 627-635.

Hansen, J.W. 1996. Is agricultural sustainability a useful concept? *Agricultural Systems*. 50: 117-143.

Heinen, J.T. 1994. Emerging, diverging and converging paradigms on sustainable development. *International Journal of Sustainable Development, World Ecology* 1: 22-33.

Hermanidez, G. Nijkamp, P. 1998. Multicriteria evaluation of sustainable agricultural land use: A case study of Lesvos. In: Beinat E. and Nijkamp. P. (Eds.). *Multicriteria Analysis for Land Use Management*. Kluwer Academic Publishers, Dordrecht.

Ikerd, J.E. 1993. The need for a systems approach to sustainable agriculture. *Agricultural Ecosystems and Environment* 46: 147-160.

ISO (International Standardization Organization). 1995. *Environmental Ecolabelling: guiding principles, practices and criteria for multiple criteria-based practitioner programs. Guides for Certification Procedures*. Geneva, Switzerland.

ITTO (International Timber Trade Organization). 1992. *Criteria for the measurement of sustainable tropical forest management*. Yokohama, Japan.

IUFRO (International Union of Forestry Research Organization). 1997. *Symposium on sustainable Management of Small Scale Forestry*. Murashima, Japan. September 8-13, 1997.

Lal, R. 1991. Soil structure and sustainability. *Journal of Sustainable Agriculture*: 67-91.

Lamnerts V.B., E.M. and Blom, F. 1997. *Hierarchical framework for the formulation of sustainable forest management standards*. Netherlands, Veeman Druckers. 82 p.

Mebratu, D. 1998. Sustainability and sustainable development: historical and conceptual review. *Environmental Impact Assessment Reviews*: 493-520.

Mendoza, G. A. 1997. An introduction to the analytic hierarchy process: Theory and applications to natural resources management. National Annual Conference. In: Proceedings of the Joint Annual Meeting of the

American Congress on Surveying and Mapping (ACSM); American Association of Photogrammetry and Remote Sensing (ASPRS); and Resources Technology Institute (RTI). April 5-10. Seattle, WA. Pp. 130-139.

Mendoza, G.A., Prabhu, R. 2000a. Multiple criteria decision making approaches to assessing Forest sustainability using criteria and indicators: A Case Study. *Forest Ecology and Management* 131: 107-126.

Mendoza, G. A. and Prabhu, R. 2000b. Evaluating and Selecting Criteria and Indicators for Forest sustainability: A Case Study on Participatory Assessment under CBFM in the Philippines. *Journal of Environmental Science and Management* 2(2): 33-53.

Mendoza, G.A. and Prabhu, R. 2000c. Integrating multiple criteria decision making and geographic information systems for assessing sustainability of tropical forests. 2nd International Conference on Multiple Objective Decision Support Systems. Proceedings: MODSS for Managing Watersheds and Natural Resources. Brisbane, Australia. Aug. 1–6. 1999. (In Press).

Mendoza, G.A. and Prabhu, R. 2000d. Multiple criteria analysis for assessing criteria and indicators in sustainable forest management: A case study on participatory decision making in a Kalimantan forest. *Environmental Management* 26(6): 659-673.

Mendoza, G.A., Hartanto, H., Villanueva, T. and Prabhu, R. 2001. Multicriteria and Critical Threshold Value Analyses in Assessing Sustainable Forestry: A Case Study. *Journal of Sustainable Forestry* (In press).

Mendoza, G.A. and Prabhu, R. 2001. A fuzzy analytic hierarchy process for assessing biodiversity conservation. In: Schmoldt, D., Kangas, J., Mendoza, G. and Pesonen, M. (Eds.). *The Analytic Hierarchy Proces in Natural Resources and Environmental Decision Making*. Kluwer Academic Publishers. Dordrecht. Pp. 219-234.

Mendoza, G.A. and Prabhu, R. 2002. Qualitative multi-criteria approach to assessing indicators of sustainable resource management. *Forest Ecology and Management* (In press).

Munda, G. 1998. Multicriteria evaluation methods in renewable management: integrated water management under drought conditions. In: Beinat, E. and Nijkamp, P. (Eds.). *Multicriteria Analysis for Land-use Management*. Kluwer Academic Publishers. Dordrecht.

Neher, D. 1992. Ecological sustainability in agricultural systems: definition and measurement. In: Olson, R.K. (Ed.) *Integrating Sustainable agriculture, Ecology, and Environmental Policy*. Haworth Press Inc. London.

Nijkamp, P. and Vreeker, R. 1999. Sustainability assessment of development scenarios: methodology and application to Thailand. *Ecological Economics* 33: 7-27.

Nijkamp, P. and Ouwersloot, H. 1997. A decision support system for regional sustainable development: The Flag Model. In: van der Bergh, J.C.J.M. and Hofkes, M. (Eds). *Theory and implementation of sustainable development modeling*. Dordrecht. Kluwer Academic Publishers. Dordrecht.

NRC (National Research Council Board of Agriculture). 1998. *Forested landscapes in perspective: Prospects and opportunities for sustainable management of America's nonfederal forests*. National Academy Press. Washington D.C.

van Pelt, M.J.F., Kuyvenhoven, A. and Nijkamp, P. 1995. Environmental sustainability: issues and definition and measurement. *International Journal of Environmental Pollution* 5(2/3): 204-223.

Poore, D. 1990. Sustainability in the tropical forest. *Journal of the Institute of Wood Sciences* 12(2): 103-106.

Prabhu, R., Colfer, C., Venkateswarlu, P., Tan, L.C., Soekmadi, R. and Wollenberg, E. 1996. *Testing Criteria and Indicators for the sustainable management of forests*. Phase I. Final Report. CIFOR Special Publication.

Prabhu, R., Colfer, C. and Dudley, R.G. 1998. *Guidelines for Developing, Testing and Selecting Criteria and Indicators for Sustainable Forest Management*. CIFOR Special Publication.

Prabhu, R., Ruitenbeek, H.J., Boyle, T.J.B. and Colfer, C.J.P. 2001. Between voodoo science and adaptive management: the role and research needs for indicators of sustainable forest management. In: Raison, J., Brown, A., Flinn, D. (Eds.). *Criteria and Indicators for Sustainable Forest Management*. CABI Publishing, Oxford.

Renning, K. and Wiggering, H. 1997. Steps towards indicators of sustainable development: linking economic and ecological concepts. *Ecological Economics* 20: 25-36.

Saaty, T. 1995. *Decision Making for Leaders: The Analytic Hierarchy Process in a Complex World*. RWS Publications.

Saaty, T. 1996. *MultiCriteria Decision Making: The Analytic Hierarchy Process*. University of Pittsburgh.
Sample, A. and Aplet, G.H. 1994. *Defining sustainable forestry*. Island Press. Washington, D.C. SCS (Scientific Certification Systems). 1994. The Forest Conservation Program: Programme Description and Operations Manual. California.
Schmoldt, D., Kangas, J., Mendoza, G. and Pesonen, M. (Eds.). 2001. *The Analytic Hierarchy Proces in Natural Resources and Environmental Decision Making*. Kluwer Academic Publishers. Dordrecht. 305 pp.
SGS Forestry. 1994. Assessors Handbook, Policy Document and Procedures Manual. SGS Forestry, Oxford, UK.
Spedding, C.R.W. Sustainability in animal production systems. *Journal of Animal Science* 61: 1-8.
Torquebiau, P.B. 1992. Are tropical agroforestry gardens sustainable? Agricultural Ecosystems and Environment. 41: 189-207.
Udo, H.M.J. and Cornelissen, A. 1998. Livestock in resource poor farming systems. *Agricultural Outlook* 27(4): 237-242.
Varma, V.K., Ferguson, I. and Wild, I. 2000. Decision support system for sustainable forest management. *Forest Ecology and Management* 128: 49-55.
Vavra, M. 1996. Sustainability of animal production systems: an ecological perspective. *Journal of Animal Science* 74(6): 1418-1423.
Verbruggen, H. and Kuik, O. 1991. Indicators of sustainable development: An overview. In: Kuik, O. and Verbruggen, H. (Eds.). *In search of indicators of sustainable development*. Kluwer Academic Publishers, Dordrecht.
Vogt, K.A., Larson, B., Gordon, J., Vogt, D. and Fanzeres, A. 1999. *Forest Certification: Roots, Issues, Challenges, and Benefits*. CRC Press.
Wenger, R.B., Harris, H.J., Sivanpillai, R. and Devault, D.S. 1999. A graph-theoretic analysis of relationships among ecosystem stressors. *Journal of Environmental Management* 57: 109-122.
Zadeh, L. 1965. Fuzzy sets. *Information Control*. 8: 338-353.

Chapter 5

USING GIS TO SUPPORT MULTI-OBJECTIVE DECISION MAKING IN FOREST MANAGEMENT

An experience from Ghana, West Africa

PETER A. KWAKU KYEM
Department of Geography, Central Connecticut State University, 1615 Stanley Street, New Britain, Connecticut, 06050, USA

Abstract. This chapter is about a GIS-based multiple objective evaluation method that was implemented in a community in Southern Ghana to foster collaboration between local groups competing for forest resources. The study adopts decision support procedures available in Idrisi for Windows GIS to settle a dispute over allocation of forest resources. The chapter begins with a brief discussion of decision support techniques available in GIS. Thereafter, a case study illustrating the application of GIS in a multiple objective evaluation involving conflicting objectives is described. The chapter concludes with a review of contributions that GIS made to the dispute settlement process.

Key words: collaborative forest management, dispute settlement, multi-criteria evaluation

Timo Pukkala (ed.), Multi-objective forest planning, 99–118.

1. INTRODUCTION

Intense competition among local groups for Ghana's forest resources has brought pressure on forestry officers in the country to widen the range of interests they consider before forest policies are made. Additionally, protests against land degradation and damage to non-forest timber products used by local communities have compelled the officials to integrate environmental and social concerns into forest management policies. The making of forest policy decisions has therefore become contentious and difficult. As a result, public participation has become critical for attaining peaceful and sound forest management in many communities in the country. Despite official recognition for public participation and the efforts that are being made to involve communities in forest management, the country has very little infrastructure to support public participation. A participatory approach to forest management requires a flexible and evolving process of planning and implementation. A major reorientation of existing forest management institutions and structures in the country will have to be made to allow local groups to participate in the collaborative forest management process.

In view of the fact that public participation programs have to respond to changing community needs and expectations, appropriate methods for group decision making remain a great challenge to the implementation of the collaborative forest management process. Recent adoption of GIS by forestry officials in the country has raised the hope of attaining an explicitly reasoned decision making process. The transfer of GIS into Ghana and other Sub-Saharan African countries therefore comes with the expectation that adoption and applications of the technology will create a supportive climate to facilitate open and collaborative forest management (Kyem 2001). The GIS technology can be used to gather, store, analyze and present forestry data in easily understandable formats to facilitate group decision making. The creation of databases of information on the country's forest resources can facilitate discussions that will lay foundation for long-term institution building in collaborative forest management. The technology can be utilized in resolving conflicts and to structure the forest management process to make it open, less controversial, and more iterative than before. Incorporating GIS into forest management will facilitate joint collection and processing of data on local forests. This will make it possible for foresters and local forest user groups to share the information and exchange ideas around specific forest management problems (Kyem 1997). Such meetings can create opportunities for building trust between the former opponents. The coalitions can later serve as building blocks for reaching consensus over disagreements about plans to protect and preserve the country's forests.

2. GIS AND DECISION SUPPORT

2.1 Background

Despite the potential uses and benefits that can accrue from the integration of GIS into forest management, much of the current GIS software lack specific decision support programs to facilitate open discussions and negotiation. It is fair to state that the Decision Science discipline has not found a broad audience outside the fields of Management Science and Operations Research. Nonetheless, the slow response from the GIS community to the development of decision support programs did not prevent business groups from tapping into capabilities of the technology to develop elaborate Decision Support Systems (DSS) including a variety of Expert Systems (Armstrong et. al. 1986, Diamond and Wright 1988). Even though the data gathering and processing capabilities of GIS technology make it an ideal tool for decision making, the lack of specific decision support programs compels many resource managers to use the technology more for mapping than for making direct decisions.

To be able to attract and retain the interests of forest resource managers in the adoption of GIS, the programs need to be simple and intuitively appealing. The current public view of GIS as a complex technology demanding much training and special skills tends to discourage users and restrict its adoption to a few highly trained individuals. This aura of complexity needs to be removed through the development of simple yet mathematically rigorous procedures for use by resource managers. Such procedures must then be linked to pleasant graphical interfaces that can be easily understood and used by forestry officers. Decision support procedures in GIS must also be flexible and iterative. In the real world, decision-making is rarely simple and straightforward. Rather, the decision-making process is often repetitive and complicated due partly to the absence of data or the unreliability of information. The process is often interrupted at several stages and responses are re-evaluated continuously till a final choice is made. A GIS program that is designed to support decision-making must be flexible enough to account for the changing and evolving thoughts of decision makers. The flexibility will allow for the incorporation of GIS into many different decision-making contexts.

Furthermore, many disputes in forest resource allocation entail subjective values embodied in local traditions and beliefs. Oftentimes the resolution of such value choices is more crucial to the settlement of the dispute than the analysis of empirical facts pertaining to the conflict of interests. In communities like those in Ghana and other developing countries, taboos and traditions form a greater portion of forest management institutions. It is therefore important that decision support procedures in GIS incorporate the capability to integrate data analysis with subjective value judgments of stakeholders. This is important because parties to a dispute settlement usually understand the true choices they face when conditions between the conflicting interests and human values underlying the dispute are explored and brought to bear on the decision making process (Bolan 1967).

2.2. Mathematical programming and heuristic techniques

Many DSS are based on mathematical programming techniques. It is the expectation of many GIS users that developers of the technology will adopt similar mathematical functions to create decision support programs. For several reasons, this desire has not been fully realized. Land resource allocation problems using mathematical programming models are generally solved outside GIS although some of such programs may have an interface with GIS (Wright et al. 1983). Mathematical programming is complex and requires a considerable period of training to ensure a better understanding of basic principles and functions (Carver 1991). Such procedures may therefore be understood by only a few GIS experts. With limitations on the storage and processing capabilities of the personal computer, the mathematical programming technique seem to be appropriate in instances where a small number of choice alternatives are to be addressed. In a raster GIS environment, massive data sets contained in grid cells of raster images typically create storage problems and delays in computing time (Eastman 1993). The large data volumes cause delays in the search for solution to problems. Furthermore, providing answers to problems by maximizing or minimizing some characteristics of the decision variables might not be appropriate for settling value-based conflicts that foresters typically confront in forest resource allocation. The mathematical programming technique is thus suitable in cases where maximization of variables is possible and also relevant to the problem under consideration. In view of these technical and other problems, heuristic techniques are more common in GIS software programs. Unlike mathematical programming that utilizes mathematical functions to compare variabes, heuristic techniques utilize sets of logical procedures to evaluate alternative choices (Eastman et. al. 1993). The heuristic techniques are more flexible, simpler, and hence more easily understood by resource managers with little background in mathematical computation (Cromley 1994). The technique is favored whenever complexity of the data sets exceeds computational storage of the common personal computer and also when human values underlying the decision problem require a flexible and open approach. In much of today's GIS, heuristic algorithms provide decision support via a variety of multi-critera evaluation procedures.

2.3. Multi-criteria Decision Support with GIS

Multi-criteria evaluation technique predates GIS development and as such many of the terminologies used in the field may not apply directly in GIS. Throughout this chapter therefore, the author adopts concepts and procedures offered for use with decision support programs available in Idrisi for Windows GIS. In their book GIS and Decision Making, Eastman et al. (1993) define a decision as a choice between alternatives, where the alternatives may be different land uses, actions, objects, and the like. Generally, a good choice involves evaluation of alternatives based on one or more criteria. A criterion is some basis for a decision that can be measured and evaluated. It is the evidence consisting of measurable attributes of an object that is being considered for a choice. In GIS, criteria are made up of factors and constraints. According to the authors, a factor is a criterion that enhances or detracts from the suitability of an alternative for the activity under consideration, while a constraint is a criterion that serves to limit the alternatives to be considered. In the GIS environment, constraints are represented as boolean maps where areas excluded from consideration have a zero value while areas open for consideration have a value of one. On the other hand, factors are represented as continuous scores for the activity under consideration. The procedure by which criteria are combined into a single composite index to arrive at a particular evaluation is termed a decision rule. Decision rules are generally structured in the context of a specific objective. An objective is defined as the perspective of a stakeholder that guides the structuring of a decision rule.

Multi-criteria evaluation of a management objective

In a group decision-making context such as the case study described later in this chapter, stakeholders often approach the discussion with specific perspectives (their principal objectives) that define what they would like to achieve. These primary objectives form the basis of their motivation for participating in the discussion. The objective therefore serves as a guide to the structuring of the decision making process and hence the multi-criteria and multi-objective evaluations. In the GIS-based multi-criteria evaluation adopted for this study, several criteria (factors and constraints) are used to assess each individual objective. For example, in the forest resource allocation project, each party focuses on one primary management objective (i.e., logging and forest preservation). Each of the objectives necessitated identification of a set of criteria to evaluate the relationship between the desired goal and land-based factors that carry the activities sought by the groups. In evaluating the relationship in GIS, the criteria were transformed into suitability levels regarding the objective under consideration. This was done so that the multi-criteria evaluation (or the suitability mapping technique) would allocate the resources equitably among the parties involved in the discussion.

There are currently different approaches to multi-criteria evaluation in GIS but generally, the approach involves all or several of the following steps (Kyem 2000).

1 Stakeholders (i.e., forest user groups) with similar or different perspectives on an issue are brought together to negotiate or discuss the issues at hand.

2 Criteria (attributes, or factors and constraints) that are relevant for the evaluation of the different management objectives are identified and processed in a GIS.

3 For each evaluation criterion (factor or constraint), a map that translates the attributes into a measure of suitability for the given management objective is developed.

4 Weights reflecting stakeholders' preferences regarding the evaluation criteria are generated and incorporated into the decision making process. The weights express the relative importance of each evaluation criterion in regard to the objective that is being evaluated.

5 A decision rule is applied to weigh each criterion and then calculate a single suitability map for each of the management objectives. The alternative category that maximizes the objective is chosen.

6 Having created suitability maps to represent the individual objectives, additional procedures are adopted to resolve conflicts contained in the suitability maps so that disputes can be settled.

Some of the most common techniques adopted for multi-criteria evaluation in GIS are described below.

Boolean overlay technique
The boolean overlay technique for multi-criteria evaluation requires all criteria to be converted into true/false (or 0/1) logical statements of suitability for the management objective that is being evaluated. The boolean maps represent restrictions imposed on choice alternatives. They are often created for the elimination of areas that are considered unsuitable for the human activity under consideration. The constraint maps are often summed up by some combination of intersection (logical AND) or union (logical OR) operators to produce a suitability map. The boolean technique is commonly applied in vector GIS but can also be applied in raster GIS systems.

Weighted linear combination technique
The most common method for multicriteria evaluation is the weighted linear combination (WLC) of factors and constraints (Voogd 1983). In the WLC technique, factors are evaluated as fully continuous variables rather than discrete boolean constraints. For example, if the distance from a road is considered important in a decision to locate a project, a road factor map is developed to express proximity to the road or a continuous incremental cost distance from it. The factors are then transformed and assigned some weights. Finally, the criteria are combined by means of a weighted linear combination formula as shown below.

$$S = \sum_{i=1}^{I} w_i x_i \prod_{j=1}^{J} c_j \qquad (1)$$

where S = suitability, w_i = weight of factor i, x_i = criterion score for factor i, c_j = criterion score for constraint j, I = number of factors, and J = number of constraints.

In the evaluation process, each factor is multiplied by a weight followed by a summation of the results to yield a suitability map that is then masked by a product of the constraint maps. The WLC technique is frequently applied in raster-based GIS systems. In section 3 of this chapter, we describe how the WLC technique was used in a GIS to mediate a multi-objective problem involving the allocation of forest between conflicting objectives.

Ordered weighted average technique

A more recent approach to multi-criteria evaluation is the Ordered Weighted Average (OWA) proposed by Eastman and Jiang (1996). This is a variant of the weighted linear combination technique that employs fuzzy set logic in the evaluation of criteria. Two sets of weights are applied: criterion weights that apply to specific factors and ordered weights which apply to the factors as determined by their rank ordering after the normal weighting step has been completed. Trade off between the factors is controlled by the degree of dispersion within the ordered weights. By varying the ordered weights therefore, a continuum of aggregation operators is produced. At one extreme (minimum) is the hard intersection logical AND and at the other extreme (maximum) the union operator prevails. The traditional WLC operator lies halfway between the two extremes. Unlike the WLC technique, one can profoundly change the nature of the decision rule by altering the ordered weights.

2.4. Multi-objective evaluation

The multi-criteria evaluation technique adopted for this study utilizes one or several criteria to assess parcels of land regarding individual management objectives. The technique is rarely used to resolve conflicts resulting from multiple management objectives. Although the GIS-based multi-criteria evaluation procedure (when used as a suitability mapping technique) can help identify the most desirable location for siting projects associated with given objectives, it is rarely applied to resolve conflicts arising from competing activities such as logging and forest preservation. In cases where the resolution of multiple or conflicting interests are involved, one defers to multi-objective evaluation methods. Such methods can simultaneously evaluate two or more objectives to resolve conflicts or produce an optimum allocation from among several competing objectives (Eastman et al. 1993). While there is some tradition of multi-criteria evaluation in GIS, the incorporation of multi-objective analysis in GIS is more recent and hence less developed in much of the current GIS software (Cromley, 1994). There are several classes of multi-objective evaluation techniques that could be adopted in a GIS application. They include weighting methods, ranking or prioritizing techniques and non-inferior solution methods. With the weighting technique (which form the basis of decision support procedures adopted in this study), solutions are first found for single objectives and a point associated with the optimal solution (ideal point) is established for each objective. Allocations are then carried out based on the weighted sum of the relative deviation of each criterion score from the ideal point. In a ranking method, a rank rather than a weight forms the basis of the final allocation of criterion scores. The third method is made up of sets of non-inferior solutions. In a non-inferior solution the performance measure of one objective cannot be improved without degrading the performance measure of another objective.

Many decisions in forest management are made to attain *one objective* (such as timber production) but generally, multiple objectives drive decisions in forest resource allocation. In multiple land allocation, involving for example, forest preservation and the creation of a national park, the multiple objectives complement one another. With conflicting objectives (i.e., wildlife preservation and clear-cut logging) the activities are usually in spatial competition for the use of the same plot of land. In the rest of this chapter, the author describes a GIS application in a multi-objective evaluation involving conflicting objectives. The conflict of interests revolves around the allocation of forest resources between two rival forest user groups in a community in Southern Ghana.

3. CASE STUDY: MANAGING CONFLICTS OVER FOREST RESOURCE ALLOCATION

3.1. Background to the dispute

The rain forest from which Ghana's timber and major export crops are produced is located in the southern part of the country. The forest originally covered a land area of about 8.3 million hectares, of which 1.8 million hectares was under permanent forestry and duly demarcated, surveyed and constituted as "Forest Reserves" managed by a Forest Department (now the Forest Service). About 1.37 million hectares (76%) of the reserved forests are managed as productive forest while the remaining 0.43 million hectares (24%) fulfils a protective function against the southward advance of the dry savannah vegetation. The country's forest reserves, including the Aboma forest, have recently acquired important status as a source of capital for Ghana's economic development. On the other hand, annual forest fires and an increasing reliance on the forests to meet domestic food needs have increased pressure on the forests. These have accompanied a surge in demand for fertile lands and compelled farmers to fall back on their ancestral lands preserved under forest reserves (Kyem 1997). The result has been widespread deforestation.

After years of hostilities during which coercive measures adopted by the state failed to prevent local groups from encroaching on the forest reserves, the government began forest policy reviews in the early 1990's. The reviews culminated in a major policy change in forest management. The state realized that a much surer way to ensure the protection and efficient management of the country's forest reserves was to empower communities through the creation of partnerships between foresters and local groups (CFMU 1993). To facilitate the building of collaborative forest management institutions in support of the project, the author designed a study around Idrisi for Windows GIS and implemented it in communities in the forest region. This section tells about a project that was implemented at Kofiase to settle a dispute over the allocation of forest resources in the local Aboma Forest.

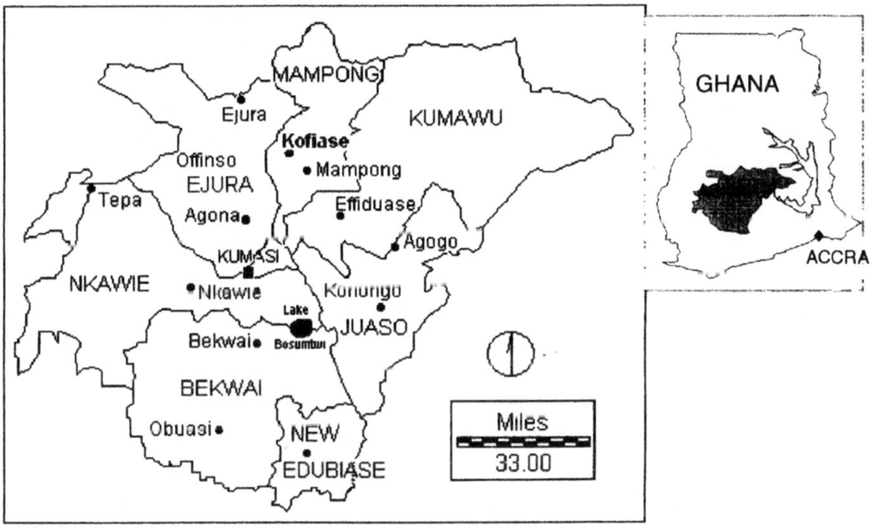

Figure 1. The study area: forest districts in the Ashanti region of Ghana

Kofiase is located fifty kilometres north east of Kumasi, the capital of the Ashanti Region, and about three hundred and seventy kilometres north of Accra, the capital of Ghana (Fig. 1). There are about 2,000 people in Kofiase and surrounding communities. The majority of people in the area are subsistence farmers who also produce cocoa as a cash crop for export. Logging in the local Aboma Forest Reserve used to be the second source of employment for the people after cocoa cultivation. The people of Kofiase and surrounding communities depend on the local forest for income, wood for fuel, protein supplements, sources of drinking water and many of their domestic needs (Falconer et al. 1994). Since the early 1980s, wild fires destroyed much of the 4,566 hectares of forest and farms in the community.

In the months leading to the implementation of the project, a logging company that held the concession to timber resources in the Aboma Forest Reserve decided to log the non-degraded portions of the forest. The decision divided the community between those inhabitants who backed the logging and others who wanted to preserve the forest and thereby protect resources that sustain the community. Soon after we arrived in the town, we formed a collaborative forest management committee and invited supporters of both sides of the dispute to a meeting in an attempt to manage the conflict with GIS. There were demands to both preserve and log the remaining forest and as such the interests of the two parties were in conflict. Both parties competed for resources within the same location in the local forest. In managing the conflict, the demand by each of the parties was treated as an independent, single objective multiple criteria evaluation problem. Results of the two single objective solutions were then used to resolve the conflicts in the multi-objective problem.

3.2. Resolving the single objective multiple criteria problems

Developing criteria for the logging objective

Separate discussions were organized for the parties to find out conditions that would fulfill their interests. The group that supported logging made the following demands:

- 350 hectares of the forest that was stocked with timber must be set aside;
- the site must be close to a road to allow for easy transportation of logs to haulage depots and processing centers in the region;
- the site must be flat to gently sloping to allow for easy logging and haulage of the timber; and
- very steep slopes or rugged terrain areas in the forest must be avoided.

Further discussions with the group resulted in the identification of criteria that were used to evaluate suitability of the forest for logging. The factors included: 1) roads that could be used to transport logs; 2) towns and villages where workers could be found or live; 3) a cut off point for slopes beyond which logging could not take place; 4) a land cover map of the partly degraded forest; and 5) a map of private cocoa farms within the forest reserve. The factors were digitized from existing maps of the forest and processed in the GIS to produce criteria for evaluating the logging objective. The final set of evaluation criteria comprised the following:

- *Roadfac* - a road factor map expressing proximity of the forest to roads within the area;
- *Townfac* - a town factor map expressing cost distance from towns and villages in the area;
- *Reliefac* - a map expressing general relief and slopes (above 30° gradient) within the reserve;
- *Logfac* - a timber factor map expressing the cost distance from a spot in the forest where some of the most valuable timber occurred in a large concentration;
- *Lancofac* - a land cover map of the reserve portraying varying degrees of logging suitability according to the nature and form of the land cover types; and
- *Farmcon* – a boolean map showing private cocoa farms that occur in the forest.

Developing criteria for the forest preservation objective
Similar discussions were held with the group that objected to logging to identify conditions that satisfied their preference for preserving portions of the Aboma Forest Reserve. The group made the following requests:

- preserve 400 hectares of the remaining forest to protect the sources of local streams;
- retain a 200 meter buffer of forest around all streams; and
- preserve portions of the forest that occurred along steep slopes to reduce erosion and consequent land degradation.

Working from the stated preferences, the group identified features that were later used as criteria to evaluate the forest preservation objective. The factors included streams, land cover types, slopes and roads occurring within the forest reserve. The features were digitized and processed in the GIS as before to produce a set of evaluation criteria that included the following:

- *Streamfac* - a map expressing distance from streams in the forest to surrounding villages;
- *Sourcefac* - a map expressing distance from sources of streams to villages in the area;
- *Forestfac* - a map expressing distance from the villages to non-timber forest products in the forest;
- *Reliefac* - a map expressing cost distance from the summits of hills in the reserve; and
- *Farmcon* - a boolean map showing private cocoa farms that occur in the forest.

Standardizing and weighting the evaluation criteria
After the criterion factors had been identified and processed, they were standardized to a numeric score ranging from 0 to 255 to facilitate their combination and comparison

within the GIS. It is obvious that there would be differences in the degree of influence that each of the evaluation criteria could have on the outcome of the multi-criteria evaluation process. For example, the loggers considered the road factor to be more important for logging than the factor expressing proximity to towns. To allow for the incorporation of these subjective preferences into the choice options that the groups were going to make, a provision was made for stakeholders to assign weights to the evaluation criteria. The weights expressed the relative importance that the parties attached to each criterion in terms of the objective under consideration. Assigning weights to factors can be carried out through ranking and other methods. In this case, a group participatory procedure in Idrisi for Windows GIS software, based on the pair-wise comparison method developed by Saaty (1977) was adopted to develop subjective weights. First, a pairwise comparison matrix was designed for the evaluation criteria developed for both the logging and forest preservation objectives (Tables 1 and 2). After that we worked across the rows from the bottom half of each matrix to determine the priority ratings by comparing the row factors against the corresponding column factors. The process was accomplished by asking decision makers to rate the factors and then provide reasons for their choices.

Table 1. Pairwise comparison matrix for the evaluation criteria (factors) of the logging objective.

	Roadfac	*Townfac*	*Logfac*	*Landcofac*	*Reliefac*
Roadfac	1				
Townfac	1/3	1			
Logfac	3	5	1		
Landcofac	1/5	1/3	1/5	1	
Reliefac	1/5	1/3	1/5	1/3	1
Weights	0.26	0.13	0.48	0.08	0.05

The pairwise ranking and weighting exercises produced lively discussions. Participants continued to probe each other's suggested rank until consensus was reached on a final rating. The exercise proved effective in helping the two parties to reflect on their demands. The matrix was eventually fed into a WEIGHT procedure in the GIS software to derive a set of weights that summed up to one (Tables 1 and 2). To produce the weights, the GIS program calculated the principal eigenvector of a square reciprocal of the pairwise comparison ratings. After each round, the program checked for internal consistency in the overall pairwise ratings. When an error was detected, the program prompted participants for a new rating and disclosed columns where the errors occurred. The weighting exercise provided an opportunity for many of the participants to express their opinions about the process.

Table 2. Pairwise comparison matrix for the evaluation criteria (factors) of the forest preservation objective.

	Streamfac	*Sourcefac*	*Forestfac*	*Reliefac*	*Landcofac*
Sourcefac	1				
Streamfac	3	1			
Forestfac	1/3	1/3	1		
Reliefac	1/5	1/7	1/3	1	
Landcofac	1/3	1/5	1/3	1/3	1
Weights	0.26	0.46	0.14	0.08	0.06

Creating single objective suitability maps

Finally, the factor scores for each objective were multiplied by their associated weights. The results were combined by means of a weighted linear combination (Equation 1).

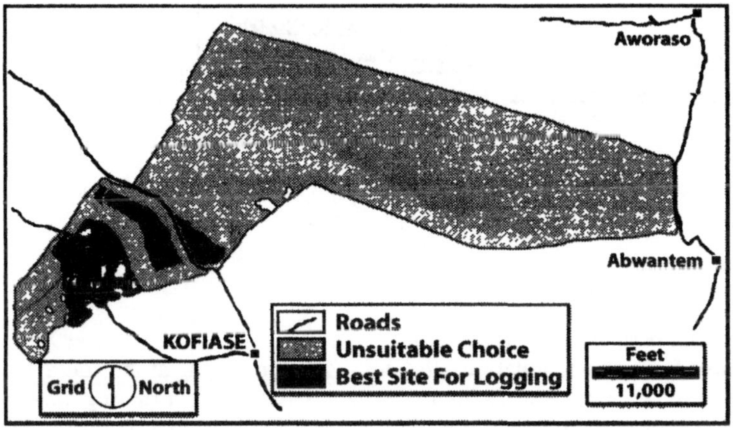

Figure 2. Aboma Forest Reserve: Best 350 hectares for logging

This resulted in separate suitability maps for the logging and forest preservation objectives. The Aboma Forest Reserve covers an area of 4,566 hectares. The forest was mapped using a 30-meter resolution data set. Each pixel therefore occupied an area of 900 m^2 of forest. Since a hectare covers 10,000 m^2, 11 pixels in the data set represented a hectare (10,000/900). Accordingly, the 350 hectares of forest demanded for logging translated into 3,889 pixels while the 400 hectares demanded for preservation required 4,444 of the same pixels. With this information, we ranked the pixels in each suitability map in an ascending order of magnitude and selected a quantity of the top ranked cells that made up area targets demanded for each objective (Figs. 2 and 3). It was clear from the two maps that the zone of maximum suitability for both objectives occurred at the southwestern portion of the Aboma Forest Reserve. Owners of private cocoa farms in the area had prevented annual fires from damaging the forest.

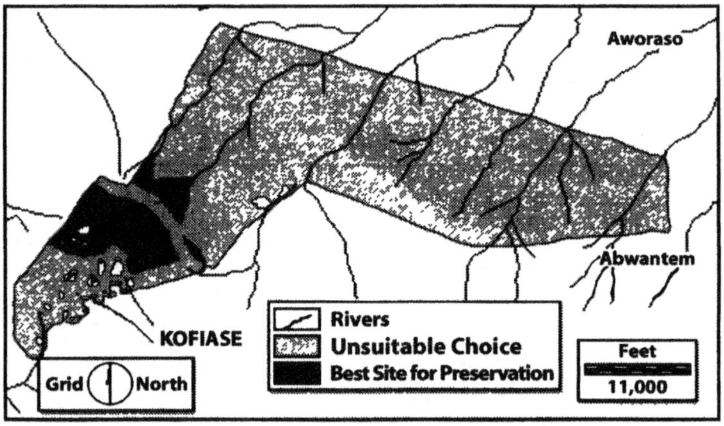

Figure 3. Aboma Forest Reserve: Best 400 hectares for preservation.

3.3. Resolving the multi-objective problem

Exploring spatial dimensions of the conflict.
With the zone of maximum suitability in both maps converging at the southwest portion of the forest, it became obvious that we had a case of conflicting objectives. As illustrated in Figure 4, if one considers a multidimensional decision space, the overlaying of two independent but conflicting decisions will result in four distinct regions. These include:

- an area selected for objective 1 only and hence non-conflicting;
- an area selected for objective 2 only and hence non-conflicting;
- a sizeable area not selected for either objective 1 or 2 because it is an unsuitable choice; and
- an area selected by both objectives 1 and 2 and hence in dispute (conflict zone).

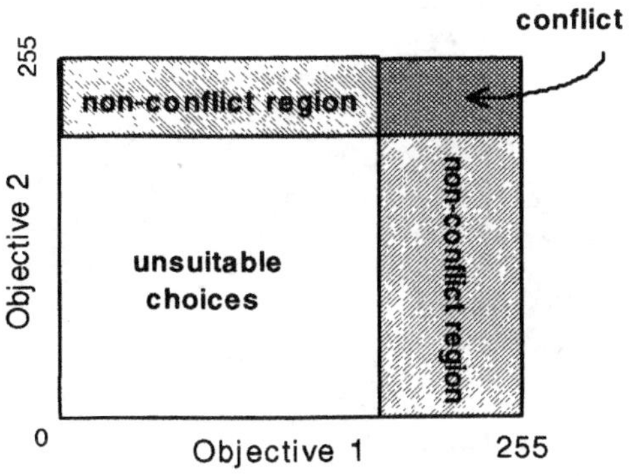

Figure 4. Allocation between two conflicting objectives

To illustrate the spatial manifestation of the conflict and thereby guide stakeholders to reflect on their demands, we cross-classified the two ranked suitability maps to create a conflict map (Fig. 5). In the map, areas in the forest that were not in dispute were separated from those areas that were jointly demanded by both parties and were therefore disputed. The map also revealed a large portion of the forest that was of little interest to the parties and hence not vital to the settlement of the dispute. The interests of the parties rather converged at the southwestern portion where the forest, as well as many non-timber forest products used by the communities had remained intact. The conflict map was used to draw the parties' attention to the spatial dimensions of their demands and to areas in the forest where their interests overlapped. The discussions prepared the parties for the difficult choices they were making to settle the dispute.

Resolving the conflict of interests

The ranked suitability maps were fed into the GIS and a multi-objective land allocation procedure (MOLA) was used to resolve conflicting cells within the conflict map. In the process, top-ranked cells in each suitability map were allocated until targets demanded by the parties were attained. This implied that some cells in one objective that happened to be marginally more suited to the other objective were traded off in exchange for less suitable cells until area targets were met. The final allocation between the two objectives is displayed in Figure 6.

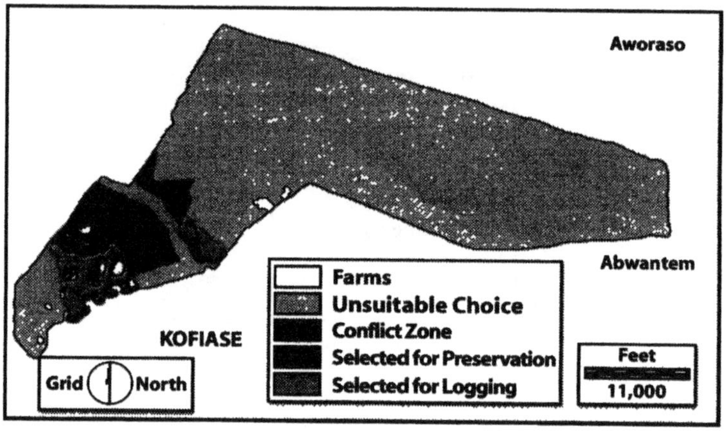

Figure 5. Aboma Forest Reserve: Conflict map

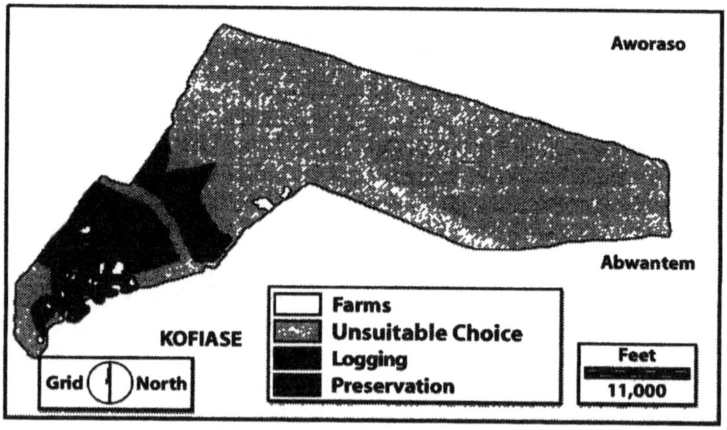

Figure 6. Aboma Forest Reserve: Final allocation map

The group which supported logging agreed to the final allocation. Many of the representatives who favored forest preservation also agreed but a handful of the latter group appeared emboldened by what they saw in maps used for the project. They became convinced that streams and other non-timber forest resources used by the community would be destroyed if logging were allowed in portions of the Aboma Forest Reserve that remained intact. The expected outcome of the dispute settlement process was realized when the parties met with the chief and elders of the town a few weeks after the GIS project. At the meeting, the final allocation map was used as a guide to delineate a zone for a downscaled logging so that the remaining forest would be preserved. The project thus succeeded in laying the foundation for amicable settlement of the dispute after several attempts had failed to reconcile differences between the two groups.

4. DISCUSSION

4.1. GIS contribution to the success of the multi-objective evaluation process

After the project, participants were interviewed about the role GIS played in the discussions. The responses from participants, including the author's observations, are summarized below.

The use of GIS to facilitate the discussions helped to curtail outbursts of emotionally charged statements that had previously characterized exchanges between the two groups. The incorporation of GIS promoted a search for factual information to support claims that were made by each party. The emphasis on evidence helped to reduce reliance on subjective judgment and speculation. As a result, a search for facts formed the basis for reaching conclusions throughout the discussions. Eventually, it was the ideas, rather than their proponents, that became the objects of discussion. The GIS approach helped to reduce direct communication, and also prevented open confrontation between the parties. In particular, the expression of dissenting views was directed toward the search for evidence (contained in maps), rather than targeted at persons. Additionally, due partly to the raising of the bar for evidential support for viewpoints, squabbles usually gave way to critical analysis and evaluation of perspectives based upon available data. Participants focused on issues that were before them and relied on facts to prove their case. An added advantage was the capability GIS brought to process data more consistently and display them in more easily understandable formats than those achieved through manual methods.

Graphic displays of the features that were subjects of the dispute (i.e., land, rivers, forest, etc.,) helped create a common image of such features among participants. The GIS maps helped to transform the mental image of the features into a common vision of the real forest resources. The concrete spatial representation of the features in the GIS maps helped to facilitate the discussions and negotiations. The technology helped to place the discussions into real contexts, spatially and otherwise. Accordingly, after each session, the parties usually understood the problems they were facing and were prepared to work out their differences.

The thorough analysis of the spatial manifestation of the conflict of interests that produced the dispute facilitated identification of hidden conflicts. For example, the conflict map revealed zones where the interests of the two parties overlapped and were hence in dispute. It also displayed areas that were not in conflict and therefore required less attention. Using the conflict map, trade-offs that could affect the settlement of the dispute were easily identified, compiled, and utilized to prompt the parties to consider implications of their demands. The map helped to focus the dispute resolution efforts on the conflict zone. By facilitating such analyses, the GIS helped to open up paths for dealing with hidden conflicts. The parties therefore became aware of the spatial implications of decisions before the final choice was made.

The adoption of the GIS strategy resulted in an open discussion with relatively equal participation occurring among stakeholders. This was possible in part because the parties were encouraged and challenged throughout the process to participate in discussions. For example, the criteria weighting process produced spirited debates that drew responses from majority of participants. Throughout the project, the GIS method required the parties to state their preferences and to speak out in support of their demands. The method also provided a mechanism for harmonizing multiple and conflicting interests. The joint data gathering and processing sessions created a congenial atmosphere that made it possible for the parties to develop some trust between them. The GIS methodology provided a formal meeting structure for the group discussions. Specific procedures were outlined and applied to the sessions. Adherence to the procedures helped participants to stay focused on issues. The method therefore laid the foundation for building blocks upon which future collaboration between the parties would be based.

4.2. Potential drawbacks of the GIS application

Notwithstanding the benefits, it is possible that the GIS application might have adversely influenced the decision making process. The powerful appeal of the technology and the brightly coloured and professional looking maps could distract stakeholders' values and obscure, rather than illuminate the true basis of their decisions. Furthermore, decision support procedures in much of today's GIS software are limited in their capability to integrate human values with factual data. Many GIS programs do not have the capability to integrate multiple perspectives and preferences into the data analysis process (Heywood et al. 1995). In particular, intangibles such as opinions, values, political considerations and other subjective preferences that underlie a dispute are easily ignored in the purely numerical analysis that takes place within a GIS.

The cost of the GIS was relatively prohibitive compared to the income of majority of people in the region. In many of the communities in the study area where we found GIS applications to be desirable, the lack of basic necessities of life made the contradictions of GIS adoption very apparent. GIS was not only foreign to the people; its impacts on institutions in the community could also not be ascertained immediately. In fact, there is currently little information about impacts that the technology exerts on social institutions in local and indigenous communities. There is little understanding of how the filtering of spatial data through foreign GIS experts confounds the meaning and understanding of issues that are of great concern to the communities. It is possible that in their excitement to adopt GIS, forest officers spearheading the implementation of the technology would ignore the adverse impacts until it is too late to address the problems.

4.3. Conclusion

The case study explained above illustrates a beneficial use of GIS in forest management. This study raises the expectation that the adoption and application of GIS in forest management would create opportunities for local groups to participate in the process. This expectation may not be realized in part because specific decision support procedures are lacking in many of the current GIS programs. There is little doubt that GIS adoption brings with it new ways of gathering, analyzing and presenting data to facilitate forest management. Adoption and applications of the system can therefore exert a significant influence on social relations at the work place, the forest policy-making process and ways of defining and mapping forests. As a technology that handles information which is undoubtedly an important basis of power in human society, GIS adoption in forest management is bound to impact established forestry institutions, structures and methods. The integration of GIS into forest management cannot therefore be problem-free. The technology brings to organizations potential benefits that are accompanied by potentially dreadful dilemmas. Additionally, creating the opportunity for Ghanaian forest officers to apply the technology in the collaborative forest management process requires computer hardware enhancement. The officers also need to learn simple procedures that can effectively involve them in discussions with local groups. This raises issues regarding technical and methodological adaptations that need to be made in current GIS software to meet the peculiar needs of resource managers and local community groups.

Notwithstanding the drawbacks of GIS applications, the current interdependence of nations, and the information medium within which GIS has evolved, make it appealing for resource managers in countries such as Ghana to adopt the technology. It is possible that if they wait, the technology will be imposed on them through some external bilateral assistance or through grants and loans for economic development. Early familiarization with GIS, however, would help the communities develop local expertise and the capacity to use the system. The people would be able to anticipate problems of GIS adoption and then develop the competence to deal with them locally. Skilled local users could initiate applications that might reform GIS applications from within their respective communities and cultural frameworks. Familiarity with the technology would offer communities the opportunity to build bases for humane and judicious uses of the technology into the system. Accordingly, an assurance for appropriate and less harmful applications of GIS would be for resource managers in developing countries such as Ghana to train and prepare for the adoption of the technology.

REFERENCES

Armstrong, M.P., Densham, P.J. and Rushton, P. 1986. Architecture for a micro-computer decision support system. In: Proceedings, of the 2nd International Symposium on Spatial Data Handling. Williamsville, New York, 120-131.

Bolan, R.S. 1967. Emerging views on planning. *American Institute of Planners* 33: 233-245.

Carver, S.J. 1991. Integrating multi-criteria evaluation with Geographic Information Systems. *International Journal for Geographic Information Systems* 5(3): 221-339.

Collaborative Forest Management Unit. 1993. Collaborative forest management news. CFMU's Newsletter 2, Kumasi, Ghana, p. 1-3.

Cromley, R. 1994. A comparison of suitability mapping and multi-objective programming techniques for performing spatial searches. *The Pennsylvania Geographer* 32: 37-46.

Diamond J.T. and Wright J.R. 1988. Design of an integrated spatial information system for multi-objective land use planning. *Environment and Planning B, Planning and Design* 15: 205-214.

Eastman, J.R. 1993 . Decision theory and GIS. Proceedings AFRICAGIS '93 Conference June, Tunis, Tunisia

Eastman, J.R., Kyem, P.A.K. and Toledano, J. 1993. *GIS and Decision Making*. Geneva, UNITAR

Eastman, J.R. 1997. *Idrisi For Windows, Version 2.0 Tutorial Exercises*, Worcester, MA, Clark Labs, Graduate School of Geography, Clark University.

Eastman, J.R., and Jiang, H. 1996. Fuzzy measures in multi-criteria evaluation. Proceedings of the Second International Symposium on Spatial Accuracy Assessment in Natural Resources and Environmental Studies, May 21-23, Fort Collins, Colorado.

Falconer, J., Wilson, E.P., Asante, P., Jampo, E.L., Acquah, S.D., Blover, E., Beeko, C., Osson, K. and Lamptey, E. 1994. Non-Timber Forest Products of Southern Ghana, O.D.A Forestry Series No. 2. Prepared by ODA on behalf of Forestry Department, Republic of Ghana.

Heywood, I., Oliver, J. and Thompson, S. 1995. Building and exploratory multi-criteria modeling environment for spatial decision support. In: P. Fisher (ed.) *Innovations in GIS 2*, London, Taylor and Francis.

Kyem, P.A.K. 1997. A GIS-based Strategy for Improving Local Community Participation in Resource Management, Allocation and Planning: The Case of Institution building for Collaborative Forest Management in Southern Ghana. Ph.D. Thesis, Graduate School of Geography, Clark University, Worcester, Massachusetts.

Kyem, P.A.K. 1999. Examining the discourse about the transfer of GIS technology to traditionally non-western Societies. *Social Science Computer Review* 17(1): 69-73.

Kyem, P.A.K. 2001. Embedding GIS applications into resource management and planning activities of local communities: A desirable innovation or a destabilizing enterprise? *Journal of Planning Education and Research* 20 (1): 176-186.

Kyem, P.A.K. 2000. A choice heuristic algorithm for managing land resource allocation problems involving multiple parties and conflicting interests. *Transactions in GIS* 5(2): 113-132.

Saaty, T.L. 1977. A scaling method for priorities in hierarchical structures. *Journal of Mathematical Psychology* 15: 234-281.

Voogd, H. 1983. *Multicriteria evaluation for Urban and Regional Planning*. Pion Ltd. London.

Wright, J., ReVelle, C. and Cohon, J. 1983. A multi-objective integer programming model for the land acquisition problem. *Regional Science and Urban Economics* 13: 31-53.

Chapter 6

HEURISTICS IN MULTI-OBJECTIVE FOREST MANAGEMENT

JOSÉ G. BORGES

Departamento de Engenharia Florestal, Instituto Superior de Agronomia, Universidade Técnica de Lisboa, Tapada da Ajuda, 1349-017 Lisboa, Portugal

HOWARD M. HOGANSON

Department of Forest Resources, University of Minnesota, 1861 Hwy 169 East, Grand Rapids, MN 55744, USA

ANDRÉ O. FALCÃO

Departamento de Engenharia Florestal, Instituto Superior de Agronomia, Universidade Técnica de Lisboa, Tapada da Ajuda, 1349-017 Lisboa, Portugal

Abstract. Heuristics have been used extensively to support forest management scheduling in the last two decades. The need for spatial definition, and the combined shortcomings of available technology and traditional mathematical programming approaches, sparked interest in alternative forest management scheduling techniques in the early 1980s. Concerns with the environmental impacts of forest management options further encouraged the development of heuristics to address adjacency relationships in harvesting decisions. More recently, heuristics have been used to target other multi-objective management issues. Namely, they have been used to provide information to help sustain both traditional forest products flows (e.g. timber and cork) and landscape structural characteristics (e.g., mosaic elements such as patch number and size, amounts of edge or interior space). In this chapter, we describe the current state of the art of heuristic application in forest management scheduling. Heuristic approaches are presented and discussed in the framework of forest management scheduling needs. Results from some heuristic research efforts are used to outline the application potential and shortcomings of these techniques.

Keywords: decision support, genetic algorithm, heuristic optimisation, multiple objectives, simulated annealing, spatial modelling, tabu search

Timo Pukkala (ed.), Multi-objective forest planning, 119–151.

1. INTRODUCTION

A heuristic may be defined as a technique that seeks good solutions at a reasonable computational cost without being able to guarantee optimality or even feasibility (Reeves 1993). The pertinence of heuristics for forest management scheduling, i.e., for land allocation and activity scheduling decisions concerning timber and non-timber outputs, was highlighted by references made in a widely used forest management textbook (Davis and Johnson 1987). This book gave a positive review of a heuristic approach developed by Hoganson and Rose (1984). Since then several authors have addressed the potential of heuristics to confront forest management scheduling problems. Why? What was the rationale for heuristic development? How do heuristics contribute to support multi-objective forest management decision making?

In the early 1980s, operations research techniques such as linear programming (LP) had increasingly replaced classical regulation methods, both at the private and public forest planning levels (e.g., Johnson and Tedder 1983). Management science and computational capacity developments justified that move. The ability to analyze alternative activities was enhanced, through the use of mathematical programming or simulation. Automation provided the means to process huge amounts of data and enabled the use of more sophisticated techniques. As a result, the potential for the definition of more sound forest plans was increased. At the same time, the formal representation of a broader range of private and public interests in the decision-making process had become possible (Borges 1999).

Yet LP was not able to convey the geographical location of forest activities. LP divisibility led to solutions where stands were split into non-spatially-referenced units to be managed differently. Further, computational constraints frequently led to the aggregation of stands into analysis areas, compounding the impact of LP divisibility. Solutions to such highly aggregated models are very sensitive to changes in assumptions and aggregation schemes (Rose 1984). In this context, the information produced by the solution may be of little value for understanding the management problem and for supporting effectively decision making. Several authors discussed these issues in the framework of the analysis of an LP-based software package used by the USDA Forest Service – FORPLAN (e.g., Dykstra 1987, Bare and Field 1987, Navon 1987). Integer programming (IP) and mixed integer programming (MIP) model building secured locational specificity but planning was still constrained by computational costs of model solving.

Thus, mathematical programming could not effectively address large-scale multi-objective problems that encompassed transportation considerations or concerns with the spatial arrangement of harvests. The computational complexity of these problems made the case for the use of heuristics. These techniques are generally more flexible and capable of addressing more complicated objective functions and constraints than exact algorithms (Reeves 1993). Moreover, given the uncertainties that derive from the large-scale attributes of the general forest management problem, approximate solutions may be adequate. As Gunn and Rai (1987) pointed out, solutions that are near optimal and near feasible may be adequate and even preferable if they can be produced with a greatly reduced solution effort, given the uncertainty about biological, technical and economic data in most forest management problems.

Further, heuristics, like exact methods, can be very useful as learning devices. They may be used to provide more insight into planning problems and suggest topics for further analysis. The distinction made by Geoffrion (1976) between the "ostensible purpose" of mathematical programming (optimization of a particular problem), and its "true purpose" (generation of information to support decision making) is illuminating. Development of heuristics for forest management scheduling thus emerged as a workable and appropriate option, particularly when multiple objectives are considered.

This chapter's objective is to present and discuss heuristic approaches to forest management scheduling[1]. This review will attempt to provide readers with a reference to most recent research and outreach work aiming at the development of heuristics and their application to multi-objective forest management scheduling problems. Results from specific heuristic research and outreach efforts will be selected to further illustrate these techniques' potential and shortcomings. This selection reflects the authors' own journey in the field of heuristic development and application. It does not reflect an evaluation of the comparative merits of all techniques referenced in the literature. Emphasis will thus be on both design and applicability of heuristics to address specific multi-objective problems rather than on a detailed listing of all techniques presented in the literature. This approach aims to provide an overview of how heuristics may be developed, and thus help to confront current challenges faced by forest managers.

Reeves (1993) refers to the explosion of interest on heuristic methods. The number of references presenting or discussing standard heuristic algorithms has increased substantially in the last few decades. Fewer works, however, emphasize heuristic design learning and understanding. This chapter attempts to fill this gap by focussing on how a modeller, facing a specific forest management problem for which a solution method is not available, might address model building and model solving (e.g. by taking advantage of specific problem structures). Yet when discussing heuristic design and application to specific management problems, a basic description of some standard heuristics is provided. Moreover, the reader is referenced to literature that describes the algorithms in more detail.

This underlying philosophy determines the chapter's organization. Thus, firstly, two techniques (an LP decomposition approach and an application of multi-attribute utility theory) will be presented to illustrate heuristic design and application. The ability to take advantage of the specific form of the multi-objective forest management scheduling problem being addressed is crucial for effective heuristic design, and will be emphasized. Based on this design, specialized optimization solution processes can sometimes be evolved to address very large and complex problems thus circumventing heuristic limitations. Both techniques presented are able to convey the geographical location of management activities at a reasonable computational cost thus overcoming limitations of exact approaches. The specialized

[1] Partial support for this research was provided by Fundação para a Ciência e Tecnologia (Project Sapiens ref. 36332/99) by POCTI and by EU fund FEDER.

LP decomposition approach and the multi-attribute approach will be further described to illustrate how heuristics may provide valuable information for multi-objective forest management. Secondly, focus will shift to a specific multi-objective management problem: adjacency. A dynamic programming based approach will be described in detail for illustration purposes. Thirdly, focus will be on multi-objective management problems that encompass both traditional product flows and spatial conditions. Several heuristics (e.g., dynamic programming based approach, simulated annealing, tabu search, genetic algorithms and sequential quenching and tempering) that may be used to address these problems will be presented. The potential for combining these heuristics with the specialized LP decomposition approach and the multi-attribute approach presented earlier for enhanced multi-objective problem representation, model solving and decision support will be discussed.

2. DESIGNING HEURISTICS TO ADDRESS MULTI-OBJECTIVE FOREST MANAGEMENT

2.1. Introduction to heuristic design

Hoganson and Rose (1984) took advantage of the specific form of general forest management scheduling problems and used an economic interpretation of the corresponding LP model to develop a specialized decomposition approach for complex problems. The integer nature of the solution of this heuristic "simulation approach" (Hoganson and Rose 1984) provides the locational specificity needed to address some multi-objective management problems. The rationale for the approach may be exposed by considering both the primal and the dual of a Model I (Johnson and Scheurman 1977) LP formulation for a simple timber management scheduling problem:

Primal model:

$$\text{Minimise } Z = \sum_{i=1}^{N}\sum_{j=1}^{M_i} c_{ij}\, x_{ij} \tag{1}$$

Subject to

$$\sum_{j=1}^{M_i} x_{ij} = A_i\,, \ \forall i \tag{2}$$

$$\sum_{i=1}^{N}\sum_{j=1}^{M_i} v_{ijpt}\, x_{ij} = V_{pt}\,, \ \forall p,t \tag{3}$$

$$x_{ij} \geq 0 \qquad \forall i,j \tag{4}$$

Dual model:

$$\text{Maximise } W = \sum_{t=1}^{T}\sum_{p=1}^{P} V_{pt} m_{pt} - \sum_{i=1}^{N} A_i a_i \qquad (5)$$

Subject to

$$\sum_{t=1}^{T}\sum_{p=1}^{P} v_{ijpt} m_{pt} - a_i \leq c_{ij} \qquad \forall i, j \qquad (6)$$

a_i unsigned $\qquad \forall i \qquad$ (7)

m_{pt} unsigned $\qquad \forall p,t \qquad$ (8)

where N = number of stands; M_i = number of options for stand i; P = number of products; T = number of planning periods; x_{ij} = area in hectares of stand i assigned to option j; c_{ij} = discounted cost for assigning an area unit (e.g. hectare) of stand i to option j including the value of ending inventory; v_{ijpt} = volume per area unit of product p in period t for stand i if option j is followed; m_{pt} = dual variable associated with the output level of product p in period t (Equation 3); a_i = negative of the dual variable associated with the stand i; A_i = area of stand i; and V_{pt} = target level of product p in period t.

In the primal model, Equation 1 expresses the management objective of minimizing management costs. Equation 2 defines the area of each stand. Equation 3 forces the desired output levels to be achieved in each period. Equations 5 to 8 represent the dual model for this problem. The primal and dual model structures show that the solution to this problem is complicated by the forest-wide timber flow constraints. These complicating constraints may be absorbed into the objective function of the dual problem. By assigning numerical values to the corresponding dual variables one is left with simple stand-level sub-problems that may be solved with standard economic cash flow analysis as the dual problem becomes:

$$\text{Minimise } W = \sum_{i=1}^{N} A_i a_i \qquad (9)$$

Subject to

$$a_i \geq \sum_{t=1}^{T}\sum_{p=1}^{P} v_{ijpt} m_{pt} - c_{ij} \qquad \forall i, j \qquad (10)$$

a_i unsigned $\qquad \forall i \qquad$ (11)

Hauer (1992) demonstrated that this solution approach is a form of Lagrangian relaxation (Fisher 1981) with specialized heuristic techniques to search for the values of the dual variables. The solution process may start with the input of an arbitrary vector of shadow prices (dual variables) associated with the forest-wide constraints (Fig. 1). These shadow prices may be interpreted as the marginal costs of providing the required timber flows. They contribute to the solutions to the stand-level sub-problems: the assignment of the optimal prescription (given those marginal costs) for each stand. They are further used to measure the impact of these stand-level solutions upon the forest-wide target product flows. The economic interpretation of the dual variables additionally enabled the development of efficient heuristic techniques to search for their values using an iterative process that stops when deviations from the target flows are within tolerable limits (Hoganson and Rose 1984). The approach thus obtains optimal near feasible solutions (Hauer 1992).

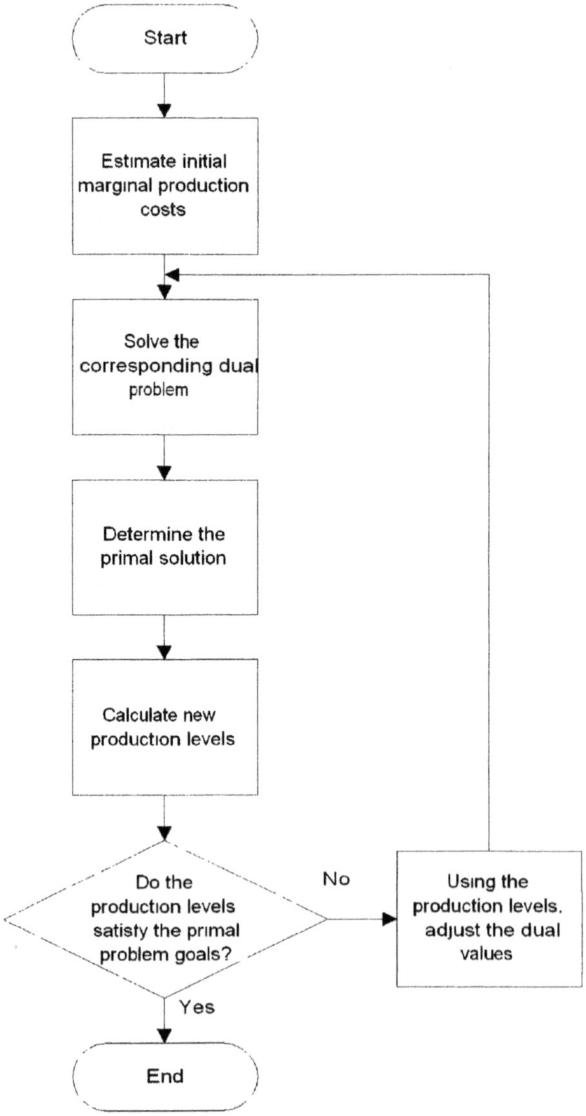

Figure 1. Flowchart of linear programming decomposition approach.

Hoganson and McDill (1993), Hoganson and Kapple (1995), Rose and Husain (1998), Hoganson et al. (1998) and Hoganson (1999) reported research and application results that show that this approach can address effectively very large multi-objective forest management problems while conveying the geographical location of management options. For example, it has been used to integrate economic concerns with multi-product target flows and environmental concerns with the distribution of forest area according to combinations of cover type and age class. Hoganson and Kapple (1991) further expanded the approach to recognize market interactions when multiple timber markets share a common timber production area. The expanded technique solved the problem of integrating harvest timing decisions and wood shipping decisions by utilizing maps of wood procurement zones. An optimal shipping schedule is identified for each stand by simply referencing its location on maps that define procurement zones for each market and product combination. The boundaries of these zones are defined by the dual variables associated with the target flow constraints and change as the approach iteratively adjusts their values. Computational savings are substantial as there is no need to enumerate all combinations of possible road building and shipping options and harvest timing options when defining the decision variables (Hoganson et al. 1998). Recently, several additional heuristics have been suggested for addressing road construction in tactical planning. Clark et al. (2000) used a first stage greedy algorithm to identify an initial solution and two solution improvement stages using interchange methods. Richards and Gunn (2000) used graph structures to formulate spatial restrictions on clearcut size and location. A tabu search approach is used to drive the search process.

The specialized LP decomposition approach has been used successfully in several research and outreach projects in Minnesota and in Canada (e.g., the Minnesota Generic Environmental Impact Statement on Timber Harvesting and Management (Jaakko Poyry Consulting, Inc. 1994)). The value of the economic information provided by this approach has been highlighted in the forestry literature (e.g., Paredes and Brodie 1988 and 1989, Jonsson et al. 1993). Recently, Strange et al. (1999) also used the LP dual value of multiple use outputs in the framework of an approach to evaluate management alternatives in multiple-use forestry. Like LP, the technique developed by Hoganson and Rose (1984) provides estimates of the value of all management options for each stand thus enabling trade-off analysis and providing valuable information for plan implementation. In this respect, it compares favourably to random search and to intelligent heuristic optimization techniques that may also be used to meet locational specificity. Still, like LP, the decomposition approach integrates the interrelations between all stand-level management options in model solving. Thus, if meeting exactly the objectives' target values is not a concern, the approach has the further advantage of producing optimal solutions to the relaxed problems. Heuristic random search approaches that depend on the way stands are sequenced for analysis are clearly inferior. The specialized decomposition approach may be linked with other techniques to address concerns with the spatial arrangements of management activities, as will be described later in this chapter.

Pukkala and Kangas (1993) developed another original heuristic approach that conveys the geographical location of management activities. In this case, emphasis was on combining the use of multi-attribute decision methods and numerical heuristic optimization techniques. The approach may be described as follows (Fig. 2). In the first step, a multi-attribute utility function is estimated (e.g., using similar techniques as in the Analytic Hierarchic Process (AHP)). For example, through graphical interfaces, pairwise comparisons (Saaty 1997 1980) may be used to estimate both the relative importance of the objectives and the partial utility (sub-priority) functions for each objective.

$$U = \sum_{i=1}^{k} r_i u_i(q_i) \qquad \forall i \tag{12}$$

where U = total utility; k = number of objectives; r_i = relative importance of objective i; u_i = partial utility function of objective i; and q_i = value of the objective variable i.

In the second step, one management option is assigned randomly to each stand and the total utility is computed. After that, the technique looks up one stand at a time and checks whether swapping management options may lead to an increase of total utility. The look up stops when no option may be found that might increase utility (Pukkala and Kangas 1993). The second step is repeated several times and the solution with the best utility value is selected as the optimal (Kangas 1999). This component of the approach thus combines random and systematic search components.

Pukkala et al. (1995), Kangas et al. (1996), Kangas and Pukkala (1996), Pukkala and Kangas (1996), Kangas (1999) and Pykäläinen (2000) reported research and application results that show that this approach can effectively address multi-objective forest management problems while conveying the geographical location of management options. For example, it has been used in Finland to address interactive planning of private non-industrial forestry; public participation in forest management scheduling; risk and attitude toward risk in forest planning; and biodiversity, recreation and aesthetics considerations in forest management scheduling. It may also take into account spatial objectives, as will be described later in this chapter.

This heuristic has been specially developed for tactical forest planning at the area or forest holding level (Kangas 1999). The use of a multi-attribute method such as AHP allows for the explicit integration of decision-makers' preferences in the process of decision analysis. Yet AHP requires that the matrix for pairwise comparisons does not exceed a maximum size of 10 (Saaty 1980) and this may be a severe constraint when complex large-scale problems are considered. In some cases, the possibility of estimating objectives' relative importance and partial utility functions, based on expertise and direct numerical valuation, may help circumvent that limitation. Nevertheless as Kangas (1999) pointed out, the flexibility of the approach, regarding the setting of objectives and the taking of subjective preferences into account for decision analysis, makes it particularly useful for customer-oriented planning of private forestry. Solutions by this technique were reported to be near-optimal (Kangas 1999).

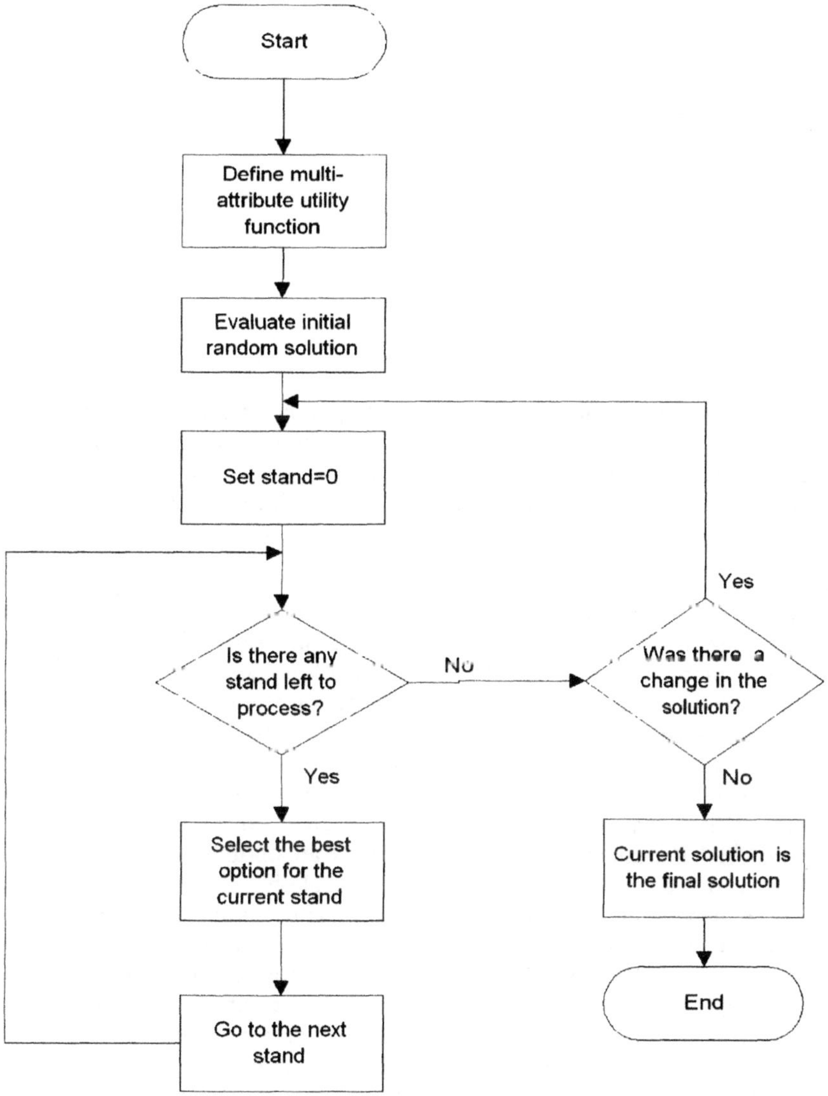

Figure 2. Flowchart of Pukkala and Kangas (1993) heuristic.

2.2. The adjacency problem

Concerns with the environmental impacts of timber production have often been formalized in legal restrictions on clearcut size. As a consequence, environmental objectives have often been addressed within the framework of the forest management scheduling adjacency problem. Borges et al. (1999) described it as:

$$\text{Maximise } Z = \sum_{i=1}^{N} \sum_{j=1}^{M_i} c_{ij} x_{ij} \quad (13)$$

Subject to

$$\sum_{j=1}^{M_i} x_{ij} = 1, \ \forall i \quad (14)$$

$$x_{ij} + x_{km} \leq 1 \qquad \forall i, j, \forall k \in A_i, \forall m \in B_{i_j k} \quad (15)$$

$$x_{ij} = 1 \vee x_{ij} = 0 \qquad \forall i, j \quad (16)$$

where N = number of stands; M_i = number of options for stand i; x_{ij} = binary variable that is set equal to 1 if option j is chosen for stand i and to 0 otherwise; c_{ij} = net present value associated with option j for stand i. It includes the value of the ending inventory; A_i = set of all stands that have an adjacency relationship with stand i; $B_{i_j k}$ = set of options for stand k that have an adjacency conflict with option j for stand i.

Equation 13 expresses the management objective of maximizing the forest net present value. Equations 14 ensure that one and only one option is assigned to each stand. Equations 15 guarantee that there are no adjacency violations, i.e., that if option j is chosen for stand i, no conflicting option is chosen for any stand adjacent to i. Equations 16 express the binary requirement on the decision variables.

This is a multi-objective forest management scheduling problem that considers both financial and environmental goals. The former are explicit in the objective function. The latter are implicit in the adjacency constraints. Moreover, it is a combinatorial optimization problem, i.e., it is an optimization problem in which decision variables are discrete and where the solution is a set of integers (Reeves 1993). In general, this combinatorial optimization problem involves very large numbers of variables and constraints. Specific formulation methods may help overcome or better address the integer program requirements (Murray and Church 1996, Snyder and ReVelle 1997, McDill and Braze 2000). Huge gains have been made in the last few years in optimization software packages (e.g., ILOG 1997). These gains are a credit to more than just faster computers. Heuristic techniques are being utilized directly in optimization packages to help in such key aspects of the solution process as (1) finding initial feasible solutions, (2) finding initial bounds on optimal solutions, (3) selecting branches to search in a standard branch and bound technique and (4) selecting the specific mathematical programming strategy to use for solving a specific mixed integer problem. Current optimization packages are also user friendly, with the potential to link directly with software developed for specific problems or heuristics. In practice much larger problems can now be addressed with exact techniques. If the heuristic techniques need not find optimal solutions it is unreasonable to assume that exact techniques must also meet such high standards. McDill and Braze (2001) emphasized that exact models are likely to find solutions to the adjacency problem very close to optimal solutions early in their solution process. In the future, differences between heuristic and exact approaches may not be so well defined, as both have the potential to be used in combination.

Yet the problem size of exact methods frequently has been a limiting factor in finding a timely solution with reasonable computational requirements (Murray and Church 1995, Weintraub et al. 2000). Heuristics have thus been widely used to address the forest management scheduling adjacency problem (e.g., O'Hara et al. 1989, Clements et al. 1990, Nelson and Brodie 1990, Nelson et al. 1991, Lockwood and Moore 1993, Weintraub et al. 1994, Murray and Church 1995, Tarp and Helles 1997, Hochbaum and Pathria 1997, Barret et al. 1998, Brumelle et al. 1998, Boston and Bettinger 1999, Bettinger et al. 1999). In the most frequently used heuristic – random search – stands are scheduled sequentially, and spatial implications associated with stands scheduled earlier eliminate options for stands scheduled later. This approach leaves open the question of optimal performance and the need for considering other forest-wide constraints is generally ignored or greatly simplified. Moreover, the computational burden of using random search may also become critical when longer planning horizons are considered. Sample size is often sacrificed to reduce computational costs (Weintraub et al. 2000). Other heuristics proposed for solving this combinatorial problem include, for example, tabu search (e.g., Murray and Church 1995, Brumelle et al. 1998, Boston and Bettinger 1999, Bettinger et al. 1999) and simulated annealing (e.g., Lockwood and Moore 1993, Dahlin and Sallnas 1993, Murray and Church 1995, Tarp and Helles 1997, Boston and Bettinger 1999, Van Deusen 1999).

Addressing adjacency concerns with dynamic programming

Dynamic programming (DP) techniques, when applicable, may be more efficient than LP or integer programming. This efficiency is critical when the need to expand the model to recognize additional details is anticipated. Hoganson and Borges (1998) have thus designed and presented a DP approach that combines characteristics of both exact and heuristic methods to address the combinatorial nature of the adjacency problem. Borges et al. (1999) extended the approach to consider forests with stands of irregular size and shape. DP may be used to find exact solutions to the adjacency problem. In the proposed DP formulation, stands define the stages (Fig. 3). States (nodes) in a given stage characterize options implied for adjacent stands addressed earlier in the DP network. Thus the number of state dimensions may be variable. Yet the formulation does not grow in size like a decision tree, because state descriptors are no longer needed for those stands represented earlier in the network that do not have adjacency relations with stands represented in a later stage. Arcs (arrows in Fig. 3) in the DP network represent management options for the associated stand (stage). Returns associated with an arc are those from the corresponding stand management option. Penalties are associated with arcs that violate adjacency constraints.

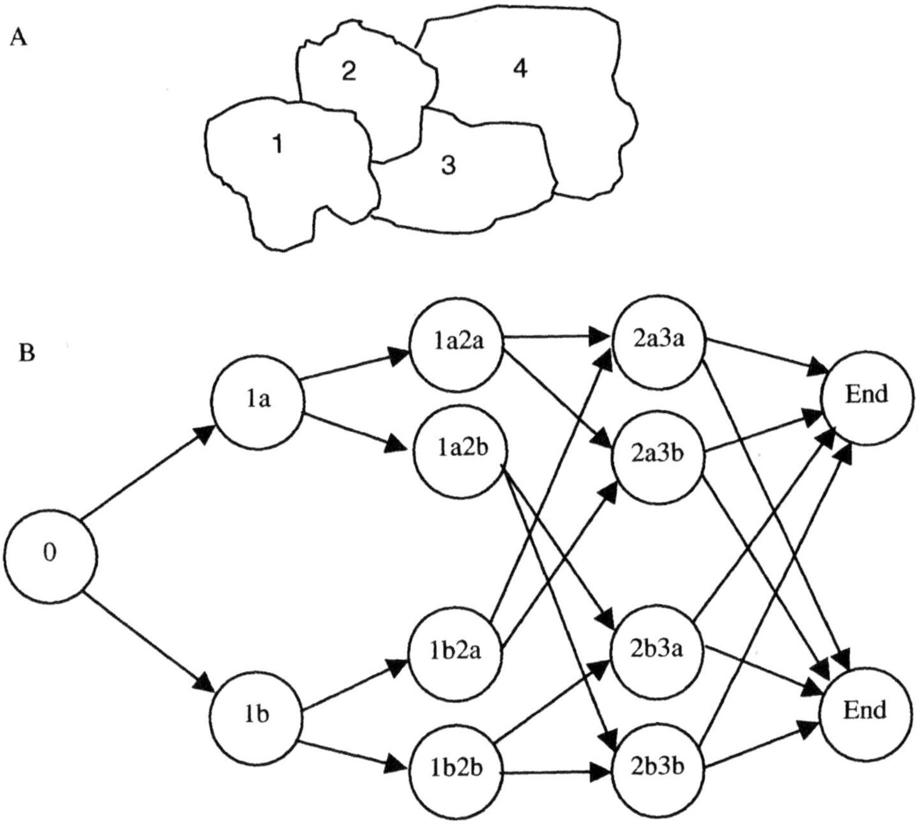

Figure 3. A: a simplified forest with 4 stands. B: the dynamic programming network generated for the above forest, processed according to the relative number of each stand. Two management alternatives are considered: a) "clearcut" and b) "do nothing".

The recursive nature of DP links the values of all stands represented in the network. In the case of many examples presented in the literature to illustrate the use of other heuristics, DP might be used to get the optimal solution as, in general, a single DP network may include a large number of stands. Borges et al. (1999) presented a mathematical description of this DP formulation:

Find $F_N^*(S_N)$ (17)

where,

$$F_i^* = \max_{X_{S_i}} \left[R(S_{i-1}, D_i) + F_{i-1}^*(S_{i-1}) \right] \tag{18}$$

$$F_0(S_0) = 0 \tag{19}$$

$$R(S_{i-1}, D_i) = NPV_{D_i} - P_{S_{i-1}, D_i} \quad (20)$$

$$S_i = S_{i-1} \theta \; D_i \quad (21)$$

where i = stage number. The total number of stages (and of stands) is equal to N; S_i = node at stage i (each node is defined by a set of state variable values that correspond to a unique set of option types for stands that still have adjacency relations with stands to be represented in a later stage; the former define the set of stands "along the front" at stage i); S_0 = initial node in the network; S_N = final node in the network; X_{S_i} = arc arriving at node S_i (it has associated with it a unique combination of D_i and S_{i-1} which is defined by equation set 21); D_i = decision made at stage i, that is choice of a option type for stand at stage i; NPV_{D_i} = net present value associated with selecting option type D_i for stand at stage i; and P_{S_{i-1},D_i} = penalty associated with option type D_i for stand at stage i, if there is an adjacency conflict with any option type in node S_{i-1}.

Equation 17 expresses the management objective of maximizing the forest net present value. Equations 18 correspond to the DP recursive formula. They compute the value of node S_i at stage i by adding current decision's financial returns to the value of the best path up to nodes S_{i-1} to which the former is connected in the network by arcs X_{S_i}. Each X_{S_i} implies a specific D_i. Equation 19 is the boundary condition. Equations 20 correspond to the DP return functions. They compute the financial impact of selecting option type D_i for stand at stage i and thus provide the value of feasible arcs in the network. Penalty P_{S_{i-1},D_i} guarantees that there are no violations of adjacency constraints as infeasible arcs in the network are assigned a very low return. Finally, Equation 21 corresponds to the DP transition function. It represents the impact of assigning stage i to a specific stand on the set of state variables defining node S_{i-1} and thus determines the set of state variables defining node S_i. The values of state variables in a given node correspond to the option types selected for all stands in the current front.

Generally, computational constraints of large problems will preclude the possibility of finding an optimal solution to the adjacency problem using the DP alone. The dimension of the DP is the limiting factor precluding the possibility of including all forest stands in one single DP network. Hoganson and Borges (1998) and Borges et al. (1999) thus developed a decomposition procedure for a large forest management scheduling adjacency problem. Typically, when decomposing the master problem, one would want to minimize the number of sub-problems generated; if only one sub-problem could be used, optimality would be guaranteed. The possibility of increasing sub-problem size relies on the ability to design the DP network so as to assign stages to stands such that the number of stands along the front does not grow unnecessarily large and exceed practical DP dimension limits. Therefore, a decomposition approach was developed to simultaneously define the sub-problems and the order in which stands are to be sequenced in a DP network. Borges et al. (1999) summarized it in six steps:

1 Select a set of stands to potentially include in the current sub-problem according to geographical coordinates and the direction for sub-problem progression.

2 Select the stand that will be represented in the first stage of the corresponding DP network for the sub-problem, according to its geographical position and the direction for sub-problem progression. This stand will define the first front.

3 In general, in any given stage i, consider all stands adjacent to a stand along the current front that have not yet been assigned a stage number. (a) Select an adjacent stand according to geographical coordinates and the direction for sub-problem progression and assign it stage $i+1$, if the resulting number of nodes and the resulting number of state variables at the beginning of stage $i+2$ are lower than the DP dimension limits. (b) Else, select an adjacent stand that would minimize the number of nodes at the beginning of stage $i+2$ and assign it stage $i+1$. (c) If no stand bordering the front may be assigned to the next stage without violating DP dimension limits, go to step 4. (d) Repeat step 3 until all stands in the area have been assigned a stage number and then go to step 5.

4 Reduce the size of the sub-problem by eliminating at least one stand from the potential sub-problem identified in step 1. Redesign the DP network defined so far according to the new adjacency relations and go back to step 3.

5 Select stands in the current sub-problem for which the optimal sub-problem solution will be accepted as part of the solution to the master problem, according to geographical location. These stands will not be included in subsequent sub-problems.

6 Solve the DP problem to obtain an optimal solution to the spatially constrained sub-problem. Use this solution to assign option types to stands selected in step 5. This assignment constrains the space of feasible solutions of the subsequent DP network. Return to step 1 if all sub-problems have not been designed and solved.

Borges et al. (1999) and Borges and Hoganson (1999) implemented the DP approach in over 20 test cases involving 800 to 3215 stands of irregular shapes and sizes. Formulations with as many as 8 periods and 19 management option types were considered. The results showed that the DP approach greatly outperformed random search techniques and still easily outperformed the approach presented by Weintraub et al. (1994). Through the generation of modified problems for which an optimal solution might be found, even if it were computationally infeasible to find it using exact methods, Borges et al. (1999) showed that the DP approach could produce solutions within 0.01 % of the optimal solution. The DP approach was further useful to assess the need to consider adjacency constraints over a long time horizon within a harvest scheduling model (Hoganson and Borges 2000). The results showed that little might be gained by considering adjacency constraints for more than a full rotation length.

When defining the forest management scheduling adjacency problem, it is generally assumed that stand boundaries have been previously delineated. Most heuristics presented in the literature address adjacency within the Unit Restriction Model (URM) framework (Murray 1999). It is assumed that stand areas are such that there are adjacency conflicts between harvests in any two adjacent units. Yet if the area of some stands exceeds the limit on maximum opening size, solving the forest management scheduling adjacency problem involves developing polygon splitting strategies. For example, Barrett (1997) used Voronoi tessellation methods to automate the process of polygon subdivision and ensured that resulting stands were blocky and still shaped by natural vegetation boundaries, thus emulating how forest land might be classified by practitioners. Conversely, stands with an area substantially smaller than the maximum opening size may be aggregated into larger units by model solving. Murray (1999) named this approach an Area Restriction Model (ARM). For example, Barrett (1996) used connected sub-graph adjacency to allow small adjacent stands to be harvested in the same planning period.

Nevertheless, the design of stand boundaries may itself impact the ability to meet target spatial conditions. For example, it may emphasize the creation of more edge or more interior space. Moreover, combining adjacency constraints with strategies for designing stand boundaries may address some key features of a forest such as age structure, spatial heterogeneity, edges and fragmentation, thus contributing to an effective macro management of forest landscapes (Hunter 1990). When addressing spatial conditions such as type and amount of edge, it may be interesting to consider also a landscape where stands of the same size, either small or large, are never allowed to be harvested in the same planning period (Borges and Hoganson 1999). According to Hunter (1990) the heterogeneity of a forest landscape with regard to species composition, size, shape or age of stands may be the key to addressing wildlife concerns. Boyce (1995) further emphasized the importance of organizing a forested landscape with a wide array of stand characteristics in order to meet current demands and to provide a flexible basis for addressing future environmental and natural resources concerns.

In this framework, Borges and Hoganson (1999 and 2000) used the DP approach to assess the impact of both stand design and adjacency constraints on forestwide spatial conditions and timber revenues. Firstly, adjacency constraints were used to ensure that all original stands maintained their identity over the planning horizon. Secondly, strategies for stand design were devised to split the original stands in order to meet two alternative maximum opening sizes. Each strategy emphasized a specific output (e.g. amount of edge, amount of interior space, timber revenues). Thirdly, adjacency constraints were used again in the case of the forest maps resulting from the stand subdivision.

Results confirmed Barrett's (1996) assertion that the impact of adjacency on timber revenues may be overestimated by management models and that the marginal gains from improved heuristics may be outweighted by finding adequate ways to design stands. The results also showed that stand design may be used for targeting both specific landscape metrics such as edge and interior space and adjacency related timber opportunity costs. They further highlighted that the effectiveness of legal restrictions on clearcut size in confronting ecosystem management may vary according to the latter goals. In some cases, adjacency constraint policy may be counterproductive because of its impact on specific landscape metrics of interest. Such policies should be based on analysis of the specific forest ecosystem management problem. In all solutions to the adjacency problem, spatial conditions like the amount and type of edge and interior space varied substantially over time, thus also suggesting that improved heuristic formulations were needed to better address spatial objectives. The DP approach has the potential to recognize additional spatial details and to adapt to a different multi-objective framework as it will be described in the next section.

2.3. Other spatial problems

The landscape spatial structure provides the framework for ecological functioning (e.g. Franklin and Forman 1987, Bradshaw 1992). Specific spatial arrangements of landscape elements correlate both with ecological features and ecosystem sustainability and with the supply of both timber and non-timber resources. Thus, strict adjacency constraints may not be adequate to address the complexity of current multi-objective management problems as they emphasize just one measurement of the land mosaic - the maximum opening size – and yet they impact other spatial elements in the patch-corridor-matrix model (Hoganson et al. 1998, Hof and Bevers 1998, Borges and Hoganson 1999, Weintraub et al. 2000). Forest ecosystem multi-objective management may rather be operationalized by considering a wider range of landscape metrics such as opening size, old forest interior space or length of edge between openings and old forest. A representation of the interdependencies of stand-level decisions at the landscape level, in building models that focus on the achievement of specific spatial outputs, may thus contribute to better forest management decisions (Borges and Hoganson 1999 and 2000).

Integrating concerns regarding the spatial arrangement of harvest and regeneration decisions in forest scheduling models is not straightforward. Exact methods have not been able to address concerns with the impact of management options on patch size, connectivity, edge or interior space. As Murray (1999) pointed out, defining in advance all potential contiguous regions is infeasible even for small problems and yet this definition may be critical for addressing landscape-wide multiple objectives by an exact method.

Heuristics have thus been used to address spatial issues other than adjacency within a forest management scheduling model. Bettinger et al. (1996) used the heuristic presented by Sessions (1992) to define wildlife habitat corridors. Bettinger et al. (1997) used tabu search to schedule timber harvests subject to wildlife habitat quality goals. Bettinger et al. (1998) further used tabu search to address concerns with the compatibility of aquatic and commodity production goals. Öhman and Eriksson (1998) used simulated annealing to model old growth interior space. Öhman (2000) further used simulated annealing to model both edge and old growth interior space. Falcão and Borges (2000) presented preliminary results of comparisons between these techniques when addressing complex multi-objective forest management scheduling problems. Falcão and Borges (2002) further developed a new heuristic technique – sequential quenching and tempering – and compared it to standard heuristic techniques (simulated annealing, tabu search and genetic algorithms). Other integrated approaches for addressing spatial concerns will be presented later in this section for illustration purposes.

According to this chapter's emphasis on heuristic design, characteristics, assessment and comparison of standard techniques (e.g. genetic algorithms, simulated annealing and tabu search) will be presented in the framework of a specific multi-objective forest management problem. This problem encompasses timber volume flow constraints, minimum and maximum clearcut opening constraints and constraints on the minimum number of old forest patches with minimum area requirements. The same problem is used to characterize the heuristic developed by Falcão and Borges (2002). The complexity of the multi-objective landscape-wide management model may be described using concepts from graph theory (Falcão and Borges 2002):

$$\text{Max } Z = \sum_{i \in V'} \sum_{j=1}^{M_i} c_{ij} x_{ij} \tag{22}$$

Subject to

$$\sum_{j=1}^{M_i} x_{ij} = 1, \ \forall i \in V \tag{23}$$

$$\sum_{i \in V} \sum_{j=1}^{M_i} v_{ijt} x_{ij} \geq (1 - e_t) P_t, \ t = 1, \dots, T \tag{24}$$

$$\sum_{i \in V} \sum_{j=1}^{M_i} v_{ijt} x_{ij} \leq (1 + e_t) P_t, \ t = 1, \dots, T \tag{25}$$

$$\sum_{j \in M_i^{Ht}} x_{ij} = 1, \forall i \in C^t \Rightarrow \sum_{i \in C^t} a_i \leq CA_{\max}, \ \forall C^t \in VC, \ t = 1, \dots, T \tag{26}$$

$$\sum_{j \in M_i^{Ht}} x_{ij} = 1, \forall i \in C^t \Rightarrow \sum_{i \in C^t} a_i \geq CA_{\min}, \ \forall C^t \in VC, \ t = 1, \dots, T \tag{27}$$

$$\sum_{j \in M_i^{Ot}} x_{ij} = 1, \forall i \in F^t \Rightarrow \sum_{i \in F^t} a_i - OF_{\min} = D_{F^t}, \ \forall F^t \in VC, \ t = 1, \dots, T \tag{28}$$

$$D_{F^t} \geq 0 \ => \ P_{F^t} = 1, \ \forall F^t \in VC, \ t = 1, \dots, T \tag{29}$$

$$D_{F^t} < 0 \ => \ P_{F^t} = 0, \ \forall F^t \in VC, \ t = 1, \dots, T \tag{30}$$

$$\sum_{F^t \in VC} P_{F^t} \geq NP_{\min}, \ t = 1, \dots, T \tag{31}$$

$$x_{ij} = 1 \vee x_{ij} = 0, \ \ \forall i \in V, \ \ \forall j = 1, \dots, M_i \tag{32}$$

where M_i= the number of options for stand i; M_i^{Ht} = the number of options for stand i that involve a clearcut in period t; M_i^{Ot} = the number of options for stand i that qualify it as old forest in period t; T = the number of planning periods; v_{ijt} = volume yield in period t that results from assigning option j to stand i; e_t = deviation allowed from target volume level in period t; P_t = target volume level in period t; a_i = area of stand i; VC = set of all subsets of V, such that for each subset the graph induced by its nodes is connected, thus representing a contiguous region; C^t = a subset of VC such that all its nodes (stands) may be harvested in period t, thus representing a contiguous region that may be harvested as a whole in period t; F^t = a subset of VC such that all its nodes (stands) may qualify as old forest in period t, thus representing a potential contiguous old forest region; CA_{max} = maximum size of clearcut opening; CA_{min} = minimum size of clearcut opening; OF_{min} = minimum old forest patch area; D_{F^t} = bookkeeping variable to check whether contiguous region F^t may qualify as an old forest patch in period t; P_{F^t} = bookkeeping variable used to count the number of old forest patches in period t; and NP_{min} = minimum number of old forest patches.

Equation 22 expresses the management objective of maximizing the forest net present value. Equation 23 ensures that one and only one option is assigned to each stand. Equations 24 and 25 force the volume levels to meet the targets in each period; it reflects concerns with the sustainability of timber flows. Condition 26 ensures that no contiguous region that exceeds the maximum clearcut opening size is harvested in any period; it reflects environmental concerns with the impact of clearcuts. Condition 27 ensures that no contiguous region with an area below the minimum clearcut opening size is harvested in any period; it reflects concerns with operational costs of timber harvesting. Condition 28 checks whether the area of old forest region F^t is greater than or equal to the minimum old forest patch area. If the condition is met, region F^t will qualify to be an old forest patch and the bookkeeping variable D_{F^t} will be greater or equal than zero; D_{F^t} will be negative otherwise. This condition reflects concerns with wildlife habitat. Conditions 29 and 30 are used to count the number of old forest patches in period t. Equation 31 ensures that the number of regions F^t that qualify as old forest patches is greater than or equal to the minimum number of old forest patches. Equation 32 expresses the binary requirement on the variables. Logical conditions (e.g., conditions 26 to 30) might also be represented by linear constraints involving binary variables (Williams 1990).

For the purpose of heuristic design, assessment and comparison, Falcão and Borges (2000 and 2002) considered test forests with 300 to 900 stands and with areas ranging from 2976 to 6250 ha. The number of Model I type binary decision variables (Johnson and Scheurman 1977) ranged from 33000 to 220000. All constraints were annual and defined over a temporal horizon extending to thirty 1-year periods.

The representation of solutions and the definition of a solution fitness evaluation function were key elements of heuristic design and implementation. For this problem, solutions in each iteration by each heuristic – simulated annealing, genetic algorithms, tabu search and sequential quenching and tempering – were represented by a vector with as many elements as stands (Falcão and Borges 2002). The value of each vector element consisted of a pointer to the management option assigned to that stand. The management option data structure itself consisted of the description and timing of all operations and corresponding inputs and outputs. The fitness function needed to evaluate heuristic solutions in each iteration may be described as (Falcão and Borges, 2002):

$$Z = \sum_{i \in V} \sum_{j=1}^{M_i} c_{ij} x_{ij} - \theta(W_{11},...,W_{pt}...,W_{PT}, Q_{11},...,Q_{pt}...,Q_{PT}) \quad (33)$$

where V, M_i , T, x_{ij}, c_{ij} as in Equations 24 to 34; θ = penalty function; Q_{pt} = proposed value of timber production or of landscape metrics p in period t (in this program, $Q_{pt} \in \{ P_1, ..., P_T, N_{CA_{max}}, CA_{max}, N_{CA_{min}}, CA_{min}, NP_{min} \}$ with $N_{CA_{max}}$ and $N_{CA_{min}}$ standing for the number of openings in any given period with an area greater than CA_{max} or smaller than CA_{min}, respectively (both are set to zero); W_{pt} = value of timber production or of landscape metrics p in period t in the heuristic solution.

Michaelewicz (1996) presented six alternative ways to design a penalty function. Falcão and Borges (2002) used a modification of the method in which penalties are static and the penalty function took the form of a parable so that small deviations from target values were less penalized than values more distant from the targets. Its parameters were further defined in order to avoid scaling problems.

Addressing spatial concerns with simulated annealing

The simulated annealing (SA) algorithm involves a sequence of iterations each consisting of randomly changing the current solution to find a new solution within its neighbourhood (Pham and Karaboga 2000). A neighbourhood *N(x, m)* of a solution *x* is a set of solutions that can be reached from *x* by a simple move *m* (Reeves 1993). In our problem, a move involves the reassignment of a management option to a stand. In order to avoid premature convergence to a local optimum, an inferior solution may be accepted. Thus, given a solution data structure and a solution fitness evaluation function, the design of a simulated annealing search strategy involves the design of a solution acceptance function and the definition of a stopping criterion (Fig. 4). This function, also called a cooling schedule, checks whether an inferior solution may be accepted. Four parameters must be specified for that purpose (Pham and Karaboga 2000): the initial temperature, a temperature update rule, the number of iterations to be performed at each temperature and a stopping criterion. The temperature is a control parameter that is used in conjunction with the quality of the solutions to check whether an inferior solution may be accepted (Reeves 1993). The probability of accepting inferior solutions increases with the temperature and decreases with the magnitude of the inferior move (Fig. 4). As the convergence process progresses, the temperature value is decreased according to the update rule, thus decreasing the probability of accepting inferior moves. The convergence process may be stopped when the temperature reaches a minimum value.

The algorithm designed and implemented by Falcão and Borges (2002) produced an initial solution by randomly assigning a management option to each stand. Equation 33 was then used to evaluate the solution vector (Fig. 4). A new solution was generated randomly changing the management option assigned to a randomly selected stand. The algorithm checks whether an inferior solution may be accepted (Fig. 4). In order to have more control over the cooling schedule, the algorithm implementation by Falcão and Borges (2002) rather than setting the number of iterations at each temperature, further extended the approach suggested by Press et al. (1992). The temperature update rule encompassed three parameters: the number of iterations leading to better solutions, the number of iterations leading to accepted inferior solutions and the number of consecutive iterations leading to accepted inferior solutions. It further included the rate at which the temperature was reduced.

Addressing spatial concerns with genetic algorithms

A genetic algorithm (GA) is a directed random search technique to seek global optimal solutions in complex multi-dimensional spaces (Pham and Karaboga 2000). The initialization process involves the creation of a population of chromosomes (Fig. 5). The value of each gene (allele) in each initial chromosome may be either generated randomly or defined according to heuristic procedure (Reeves 1993). A sample of solutions is thus likened to a set of individual chromosomes in a population. The genes in each chromosome represent the variables in a solution vector. GA may therefore simultaneously process several points of the search space, thus providing information exchange between search directions. In addition to a solution data structure and a solution fitness evaluation function, the design of a genetic algorithm search strategy involves the specification of the chromosome population size, the design of selection, mutation and crossover schemes and the definition of a stopping criterion.

Selection aims at reproducing more copies of individuals whose fitness values are higher. Usually, chromosomes to be part of the new generation are selected randomly according to relative fitness (Michalewicz 1996). Mutation and crossover act upon the new generation thus transforming it. This transformation marks the end of a GA iteration (Fig. 5). Mutation consists of changing the values of genes (alleles) in individual chromosomes. Generally, all chromosomes are checked gene by gene and the corresponding alleles are randomly changed according to a specified rate (Pham and Karaboga 2000). Crossover consists of swapping blocks of genes between two chromosomes thus transforming both. Normally, the two chromosomes are selected randomly and the blocks are cut at randomly chosen points (Reeves 1993). Both crossover and mutation are regulated by control parameters that define both the probability of any of these operators being selected and the rate at which they will act upon a new generation (e.g. the expected number of chromosomes to be recombined by crossover).

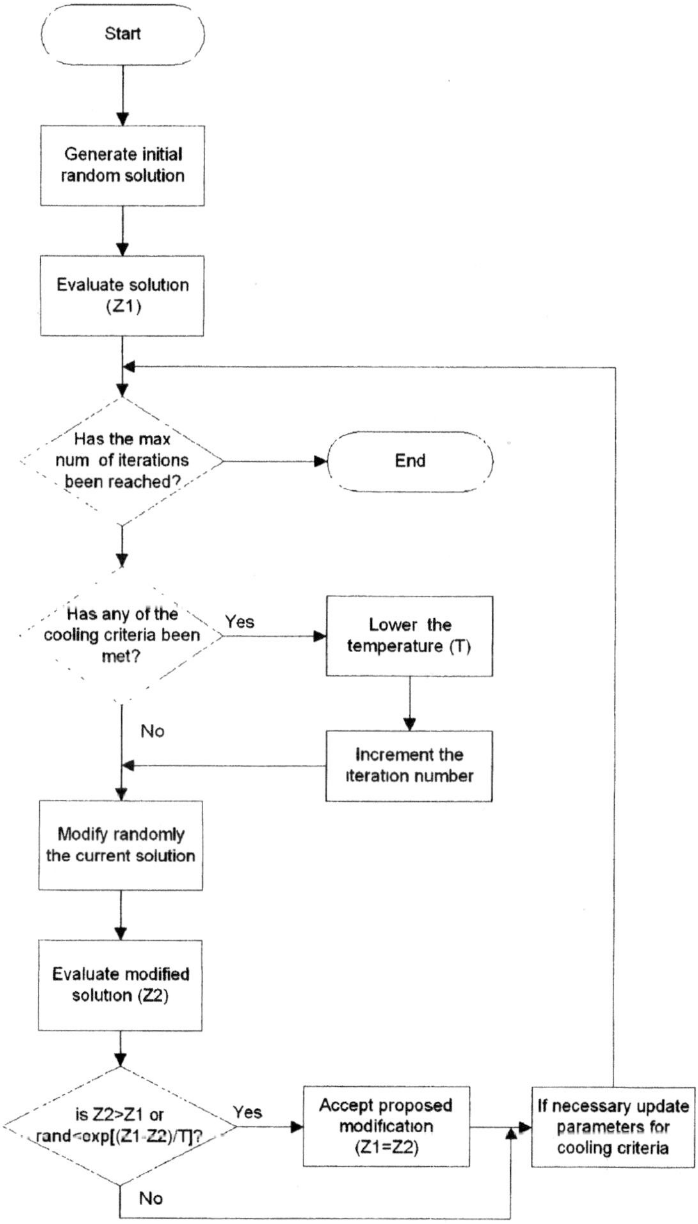

Figure 4. Flowchart of simulated annealing heuristic. Z1 and Z2: solution values before and after a proposed move; T: system "temperature"; rand: random number between 0 and 1.

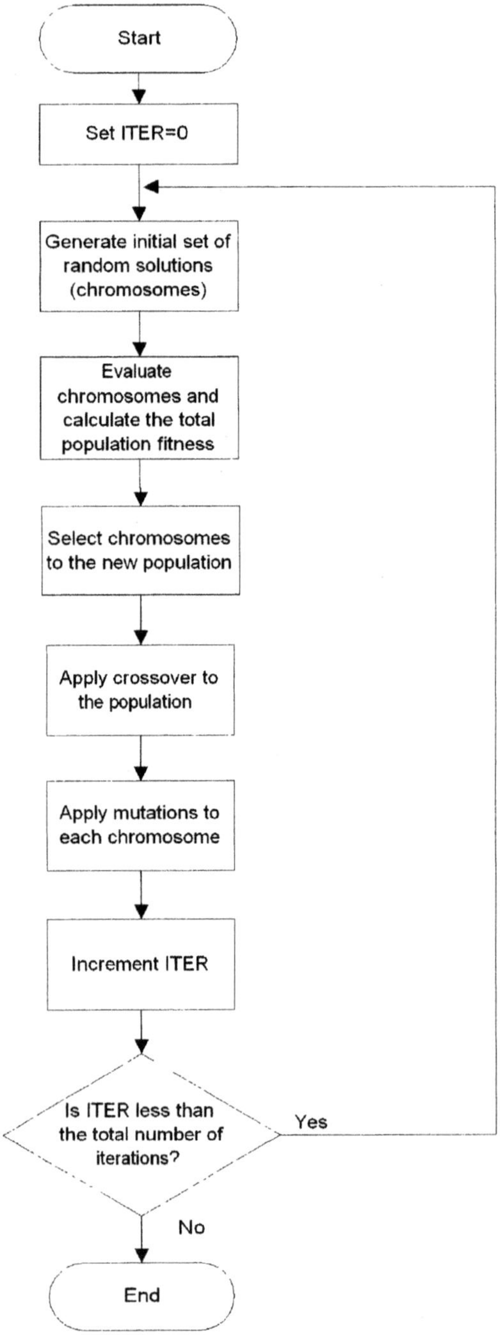

Figure 5. Generic flowchart of a genetic algorithm.

The GA implementation by Falcão and Borges (2002) created initial populations of 4 to 30 chromosomes by randomly generating the value of each gene (allele) in each initial chromosome, i.e., by randomly assigning a management option to each stand. Equation (33) was then used to evaluate the population. New solutions were generated by selecting chromosomes randomly according to relative fitness and by using crossover to recombine them. In order to avoid losing good solutions, Falcão and Borges (2002) incorporated a 'reposition of the best' strategy. The best individual in all generations so far was saved, and substituted the worst individual in a generation with a user-defined regularity. A constant crossover probability was used. Genes in chromosomes selected for crossover were organized into blocks. The number and the identification of the blocks to be exchanged were both determined randomly. Mutation procedures were further used to randomly change an allele, i.e., to randomly change the assignment of a management option to a stand. No more than two mutations were considered for each chromosome in each iteration, avoiding excess instability and the slowing of the convergence process.

Addressing spatial concerns with tabu search

Tabu search (TS) is a systematic search method that is characterized by the use of a flexible memory (Reeves 1993, Glover and Laguna 1997, Pham and Karaboga 2000). It is based on an exhaustive search of the neighbourhood solution space before accepting one change on the solution (Fig. 6). Randomization is thus de-emphasized and may be considered only for initialization purposes. In addition to a solution data structure and a solution fitness evaluation function, the design of a tabu search strategy involves the specification of two main strategies: forbidding strategy and freeing strategy. The first determined which of the solutions that result from a move from the current solution may not be accepted (Reeves 1993). These solutions define the tabu list of most recent moves kept in memory in order to avoid cycling problems and local optima. The second selects the solutions to leave the tabu list. In most cases, once the list reaches a user-defined maximum size, each newly accepted move is substituted for the oldest move in the list (Fig. 6). Yet to avoid discarding good solutions, an aspiration criterion may be used to make a tabu solution free. Intermediate and long-term memory functions may also be used to implement intensification strategies that encourage the selection of moves with features historically found good (Glover and Laguna 1997, Pham and Karaboga 2000). The definition of a stopping criterion may encompass the specification of a maximum number of iterations.

The TS implementation by Falcão and Borges (2000) tested different parameters for the forbidden strategy. The tabu list size ranged from 50 to 1,000. No aspiration criteria were considered. The number of iterations in each TS run was at least twice the number of stands.

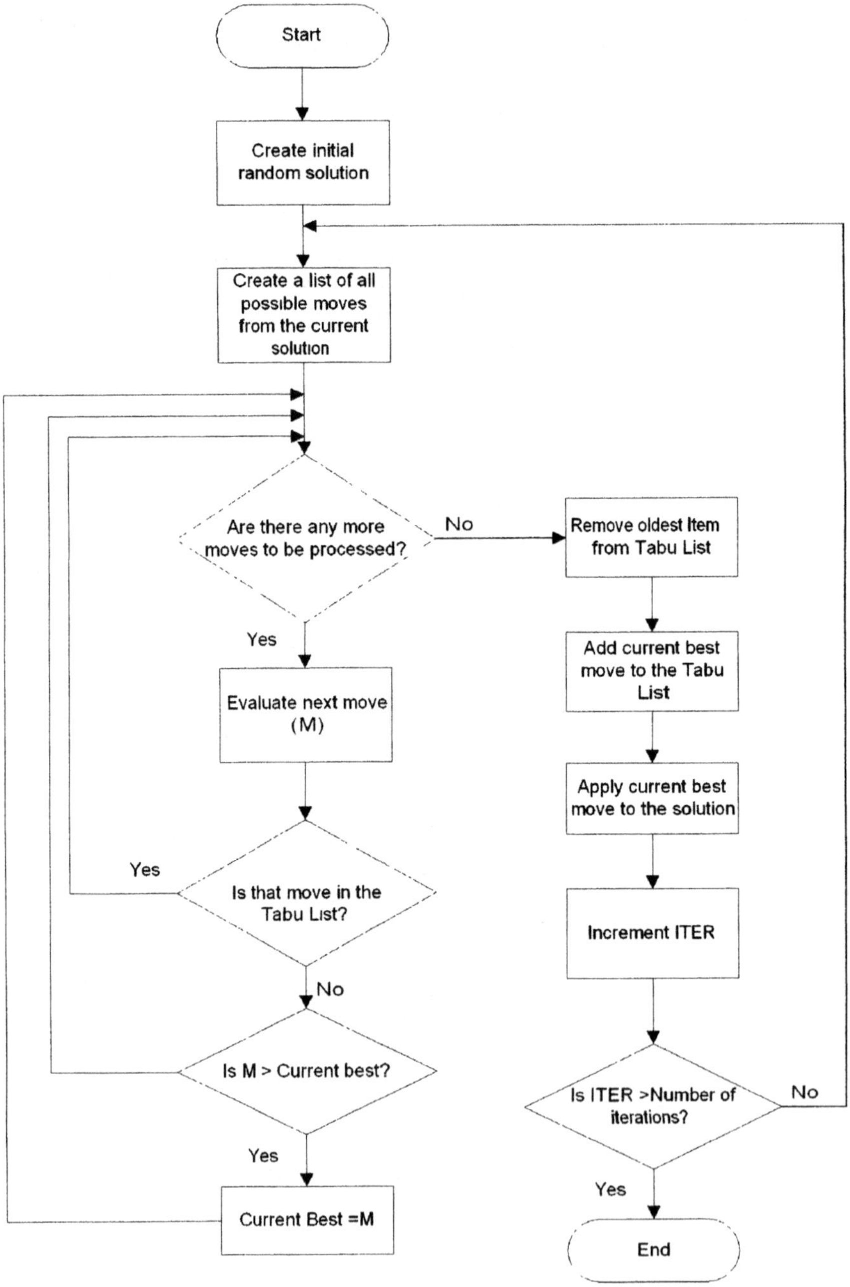

Figure 6. Flowchart of tabu search heuristic.

Addressing spatial concerns with sequential quenching and tempering

Falcão and Borges (2002) presented a new technique – sequential quenching and tempering (SQT) – for combining heuristic (random and systematic) components. The proposed algorithm produces an initial solution by randomly assigning a management option to each stand. If the maximum number of iterations has not been reached, a stand list is randomly generated for evaluation purposes. The list includes all stands in the forest. SQT searches systematically the best solution within the neighbourhood of the current solution. It checks systematically, by going through the ordered list and using Equation 33, whether assigning a management option $j \in M_i$ to stand i leads to a better solution (Fig. 7). If such an option is found, it is assigned to stand i. The same procedure is adopted for all stands in the forest (Fig. 7). In the case of TS only one move will be accepted in each iteration. In contrast, SQT may accept as many move as the number of stands. This computational efficiency may be critical for solving very large problems (Falcão and Borges 2002).

Once this systematic process is concluded, a random perturbation is applied that increases the number of potential system rearrangements. Management options are randomly assigned to a limited and randomly determined number of stands. Equation 33 is used to evaluate the new solution and the systematic search process is resumed (Fig. 7). Key control parameters are the minimum and the maximum allowable perturbation and the perturbation multiplier. The former specify the number of stands that may randomly be assigned a new management alternative. The latter is used to change the values of the minimum and maximum allowable perturbations during the convergence process.

For more details about the SA, the GA, the TS and the SQT algorithm design and implementation, the reader is referred to Falcão and Borges (2000 and 2002). Research on the discipline of combinatorial optimisation refers to potential advantages of developing hybrid techniques to solve specific problems (e.g., Reeves 1993). The results reported by Falcão and Borges (2000 and 2002) supported that assertion. Experimental results from over 1,300 test runs showed that the proposed SQT heuristic may be used to circumvent some limitations of random search approaches and that SQT is both effective and robust. Test runs further showed that if spatial constraints were relaxed, the proposed approach found better solutions than the random strategies at lower computational costs (Falcão and Borges 2002). Compared with some combinatorial optimisation heuristic techniques presented in the forestry literature, this approach extended substantially the number of binary decision variables, and of diverse environmental spatial constraints, that may be considered in the framework of an integer program.

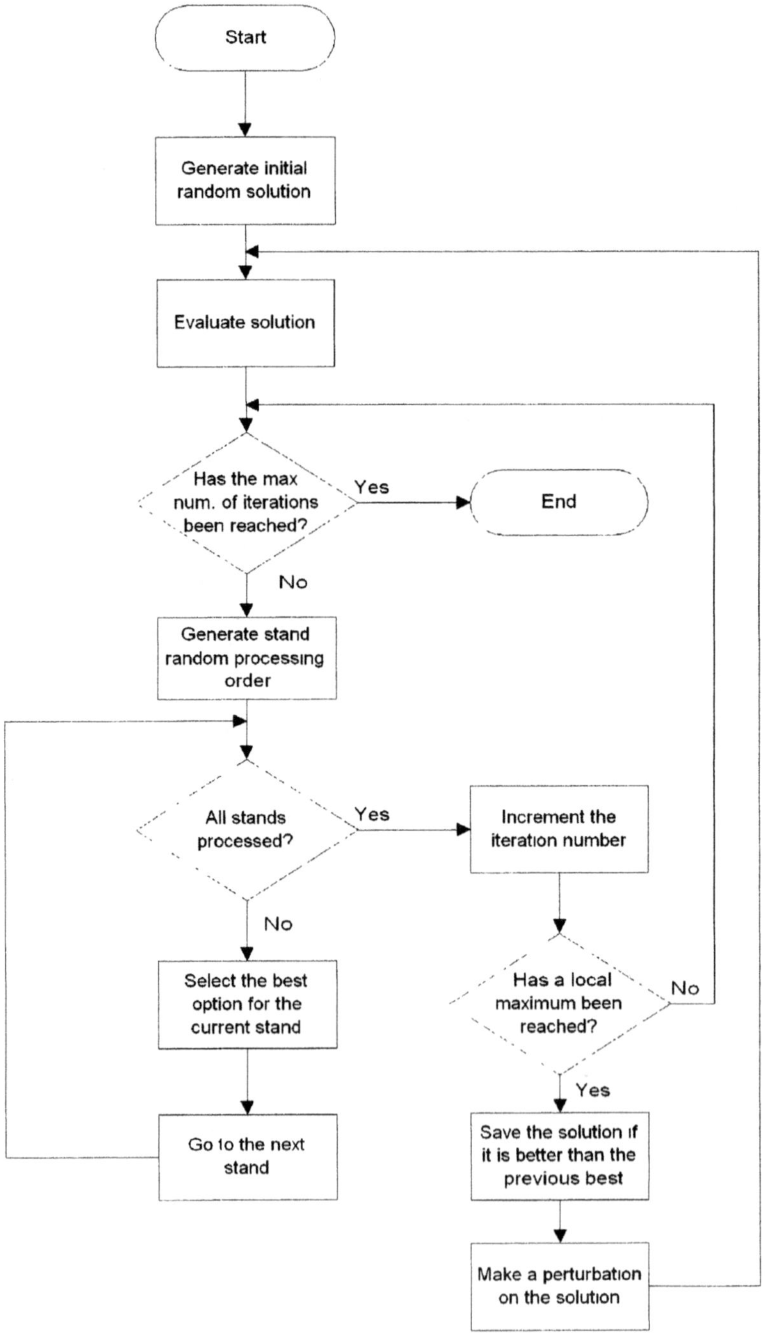

Figure 7. Flowchart of sequential quenching and tempering heuristic.

Addressing spatial concerns with other integrated approaches

To help integrate multiple objectives for timber and habitat conditions, Van Deusen (1996) proposed using a random search heuristic published by Metropolis et al. (1953). Van Deusen (1999) extended this approach to systematically adjust weights used in the objective function. The extended approach is interesting as it has features similar to both simulated annealing and the specialized LP decomposition approach described earlier. Like Hoganson and Rose (1984) Van Deusen essentially uses shadow prices to value flows for various targets for forest-wide objectives and adjusts those shadow prices at each iteration based on deviations from the targets (or acceptable levels) for the current solution. Like simulated annealing, it randomly selects new management prescriptions to consider for each analysis area and potentially accepts lower-valued solutions in this search process. Unlike simulated annealing, it does not apply a cooling process. It is also systematic in its search for improved solutions as each iteration considers a potential random change for each analysis area. The approach is also quite flexible in the types of objectives that can be included. Van Deusen (2001) has explored using it to address spatial objectives.

The DP approach designed by Hoganson and Borges (1998) and Borges et al. (1999) offers the potential for formulating landscape structure objectives in a multi-objective forest ecosystem management model. The concepts of adjacency and adjacency conflict may be expanded for that purpose. For example, adjacency may be defined in terms of distance between stands. When defining adjacency conflict, silviculture options other than clearcuts may further be considered. The forest opening may also be defined by attributes other than its size (e.g., density and species composition). Furthermore, problems may be defined in which it will be desirable to consider the relaxation of adjacency constraints. Modifications in adjacency cost assumptions make it possible to consider the impact of some violation. Rather than assigning a high cost to all arcs corresponding to combinations of management options that violate an adjacency constraint (Equation 20), benefits in fact may be assigned.

Bergmann (1999) demonstrated how the DP formulation may be expanded to model interior space to address both timber and wildlife objectives. The concept of influence zone – an area whose ecological characteristics depend on the management options selected for a specific set of stands – was introduced for that purpose. Influence zones define areas of potential interior space. Thus they are used to determine which stands must be considered as state descriptors at any given stage in a DP network. Bergmann (1999) modified the DP formulation by removing adjacency penalties from the DP return function (Equation 20) and by adding a new term $b_i(s_{i-1}, D_i)$ representing the total interior space benefits associated with selecting management option D_i for state i if the stands used as state descriptors for stage i are managed according to state S_i:

$$R(S_{i-1}, D_i) = NPV_{D_i} + b_i(s_{i-1}, D_i) \tag{34}$$

$$b_i(s_{i-1}, D_i) = \sum_{h=1}^{H} \sum_{t=1}^{T} z_i(s_{i-1}, D_i, t) \tag{35}$$

where H = number of influence zones that include stand i as the last stand addressed in the DP network; T = number of periods within the planning horizon; $z_i(s_{t-1}, D_t, t)$ = the interior space benefit for a given influence zone in period t associated with assigning option D_t to stand i if in state s_{t-1}. $z_t(s_{t-1}, D_t, t)$ is zero if the influence zone does not contribute to interior space in period t.

Bergmann (1999) implemented this DP approach formulation in two test cases involving 800 and 1272 stands classified into one of four cover type – age combinations. The number of influence zones ranged from 2416 to 4927 according to the value set for buffer width. Formulations with as many as 8 periods and 19 management option types were considered. The results showed that the DP approach did effectively help assess trade-offs between the amount of interior space and timber harvested over time.

Preliminary results from research conducted at the University of Minnesota demonstrate that through adequate formulation modification, the DP approach may also be used to effectively target other landscape metrics (e.g. amount and type of edge). Hoganson et al. (1998) further emphasized that the DP approach has the potential to work well within the specialized LP decomposition solution technique discussed earlier in this section. The DP approach could be used to solve the stand-level sub-problems during each iteration of the larger landscape-wide model. Shadow prices associated with target timber flows and landscape metrics would be adjusted according to the impact of those stand-level solutions on landscape-wide multiple objectives. Many iterations may be needed to find an adequate solution. Yet earlier in the iteration process, smaller DP formulations may be used. Tests on the DP approach suggest that very small formulations lead to good solutions. Only for the final iterations it might be desirable to consider larger DP formulations. Thus, the computational cost of the overall iteration process for the landscape-wide model might still be reasonable.

The approach presented by Pukkala and Kangas (1993) also has the potential for modelling spatial objectives. Pukkala et al. (1995) used this technique to address management concerns with scenic and recreational amenities. For that purpose, a method was developed to generate an amenity index suitable for numerical optimization. Kurttila (2001) demonstrated the use of the technique for assessing the impact of alternative spatial objective types on both timber production and spatial conditions. Habitat suitability indices (HSI) and landscape metrics were used to address wildlife habitat concerns. Spatial objectives included the proportion of a type of boundary over total boundary length, the spatial autocorrelation of HSI and a weighted mean of the stand-level HSI based on stand location. Kurttila (2001) considered test forests with 283 to 804 stands and with areas ranging from 405 to 985 ha. The planning horizons extended to three 20-year periods and to three 10-year periods in problems addressing flying squirrel and moose habitat concerns, respectively. The approach presented by Pukkala and Kangas (1993) has a modular structure that may further allow the use of other combinatorial optimisation heuristics considered adequate to address the specific multi-objective planning problem at hand.

3. CONCLUSIONS

The diversity of goods and services demanded from forest ecosystems in the present day challenges traditional management options and techniques. Current political, socioeconomic and demographic trends have led to an increasing awareness of the value of forests for wildlife survival, biodiversity protection, recreational uses and aesthetics. By emphasizing the importance of forest uses other than timber production, or even by assigning intrinsic or non-use values to ecosystems, these trends place a new challenge to forest policy makers, researchers, managers and technicians. Decision making in this area must further take into account a public opinion increasingly concerned with the social and environmental costs of current economic growth in a global and interconnected world economy. In this framework, the ability to design and solve models that may efficiently and effectively supply valuable information for decision making is crucial.

The preceding review demonstrated the potential of heuristics to address complex multi-objective forest management scheduling problems in a rapidly changing environment. Zeleny and Evans (1981) listed general conditions (e.g., inexact or limited data, unavailability or high computational cost of an exact method) under which the use of heuristics is desirable or even advantageous. Many of these conditions may be found in situations faced by forest managers today, and that helps explain the pervasiveness of heuristics in multi-objective forest resources management. Zeleny and Evans (1981) further listed several criteria that may be used for evaluating a heuristic (e.g., simplicity, computational cost, robustness, interactive ability, good stopping criteria). The heuristics referenced or presented in this review may be classified differently according to each of these criteria, yet they all were reported to meet the requirements of the multi-objective problem being addressed.

Often, the biggest bottlenecks in current multi-objective forest management are not the lack of data or technology but the lack of ingenuity in formulating adequate models (Rose 1999). Taking advantage of an understanding of the forestry problem to design adequate heuristic modeling approaches is critical to making practical application possible. This review illustrated this process by characterizing the design of several heuristics to address specific multi-objective management problems. Moreover, a single model can hardly represent all facets of a complex multi-objective planning problem. Thus it may be interesting to combine or link individual models in a logical fashion (Rose 1999). The combination may encompass both heuristic and optimization techniques. This review further illustrated how that has been accomplished by some heuristic approaches. Multi-objective forest management problems may be anticipated for which a solution technique is not currently available. The approach highlighted for heuristic design, linkage and implementation may contribute to enhanced multi-objective problem representation, model solving and decision support.

REFERENCES

Bare, B.B. and Field, R.C. 1987. An evaluation of FORPLAN from an operations research perspective. In: *Proceedings of FORPLAN symposium*, USDA, Gen. Tech. Rep. RM-140 p.133-144.

Barrett, T.M. 1996. A methodology for evaluation of the economic, forest fragmentation, and wildlife habitat effects of policies restricting size and adjacency of clearcut timber harvests. Ph.D. dissertation, University of California at Berckeley, 160 p.

Barrett, T.M. 1997. Voronoi tessellation methods to delineate harvest units for spatial forest planning. *Canadian Journal of Forest Research* 27: 903-910.

Barrett, T.M., Giless, J.K. and Davis, L.D. 1998. Economic and fragmentation effects of clearcut restrictions. *Forest Science* 44: 569-577.

Bergmann, S.S. 1999. Modeling interior space and harvest scheduling for forest management planning. M.S. Thesis, University of Minnesota, 128 p.

Bettinger, P., Boston, K. and Sessions, J. 1999. Intensifying a heuristic forest harvest scheduling search procedure with 2-opt decision choices. *Canadian Journal of Forest Research* 29: 1784-1792.

Bettinger, P., Johnson, K. and Sessions, J. 1996. Forest planning in an Oregon case study: defining the problem and attempting to meet goals with a spatial-analysis technique. *Environmental Management* 20: 565-577.

Bettinger, P., Sessions, J. and Boston, K. 1997. Using Tabu search to schedule timber harvests subject to spatial wildlife goals for big game. *Ecological Modelling* 94: 111-123.

Bettinger, P., Sessions, J., and Johnson, K. 1998. Ensuring the compatibility of aquatic and commodity production goals in eastern Oregon with a tabu search procedure. *Forest Science* 44: 96-112.

Borges, J.G. 1999. Paradigmas, tecnologias e equívocos em gestão de recursos florestais. *Revista Florestal* XII(1/2): 27-34.

Borges, J.G. and Hoganson, H.M. 1999. Assessing the impact of management unit design and adjacency constraints on forestwide spatial conditions and timber revenues. *Canadian Journal of Forest Research* 29: 1764-1774.

Borges, J.G. and Hoganson, H.M. 2000. Structuring a landscape by forestland classification and harvest scheduling spatial constraints. *Forest Ecology and Management* 130: 269-275.

Borges, J.G., Hoganson, H.M. and Rose, D.W. 1999. Combining a decomposition strategy with dynamic programming to solve spatially constrained forest management scheduling problems. *Forest Science* 45: 201-212.

Boston, K. and Bettinger, P. 1999. An analysis of Monte Carlo integer programming, simulated annealing and tabu search for solving spatial harvest scheduling problems. *Forest Science* 45: 292-301.

Boyce, S.G. 1995. *Landscape forestry*. John Wiley & Sons, Inc., New York, 239 p.

Bradshaw, F.J. 1992. Quantifying edge effect and patch size for multiple-use silviculture - a discussion paper. *Forest Ecology and Management* 48: 249-264

Brumelle, S., Granot, D., Halme, M. and Vertinsky, I. 1998. A tabu search algorithm for finding good forest harvest schedules satisfying green up constraints. *European Journal of Operations Research* 106: 408-424.

Clements, S.E., Dallain, P.L. and Jamnick, M.S. 1990. An operational, spatially constrained harvest scheduling model. *Canadian Journal of Forest Research* 20: 1438-1447.

Clark, M., Meller, R., and McDonald, T. 2000. A three stage heuristic for harvest scheduling with access road network development. *Forest Science* 46: 204-217.

Dahlin, B. and Sallnäs, O. 1993. Harvest scheduling under adjacency constraints - a case study from the Swedish sub-alpine region. *Scandinavian Journal of Forest Research* 8: 281-290.

Davis, L. and Johnson, K. 1997. *Forest Management*. McGraw-Hill Publishing Company, New York, 790 p.

Dykstra, D. P. 1987. An evaluation of FORPLAN from an operations research perspective: discussant's comments. In: *Proceedings of FORPLAN symposium*, USDA Gen. Tech. Rep. RM-140, p.145-146.

Falcão, A. and Borges, J.G. 2000. Comparison of heuristics for spatially constrained natural resources management problems. In: *Proceedings of the 2000 Symposium on Systems Analysis in Forest Resources*, September 27-30, 2000, Snowmass Village, Colorado, USA (in press).

Falcão, A. and Borges, J. G. 2002. Combining random and systematic search heuristic procedures for solving spatially constrained forest management scheduling problems. *Forest Science* 48: 608-621.

Fisher, M. 1981. The Lagrangian relaxation method for solving integer programming problems. *Management Science* 27(1): 1-18.

Franklin, J.F. and Forman, R.T. 1987. Creating landscape patterns by forest cutting: ecological consequences and principles. *Landscape Ecology* 1: 5-18.

Geoffrion, A.M. 1976. The purpose of mathematical programming is insight, not numbers. *Interfaces* 7(1): 81-92.

Glover, F. and Laguna, M. 1997. *Tabu search*. Kluwer Academic Publishers, Boston, 382 p.

Gunn, E.A. and Rai, A.K. 1987. Modeling and decomposition for planning long-term forest harvesting in an integrated industry structure. *Canadian Journal of Forest Research* 17: 1507-1518.

Hauer, G. 1992. *Timber management scheduling with multiple markets and multiple products*. M.S. Thesis, University of Minnesota, 122 p.

Hochbaum, D. and Pathria, A. 1997. Forest harvesting and minimum costs: a new approach to handling spatial constraints. *Forest Science* 43: 544-554.

Hof, J. and Bevers, M. 1998. *Spatial optimization for managed ecosystems*. Columbia University Press, New York, 258 p.

Hoganson, H.M. 1999. Management scheduling models for integrating objectives. *Revista Florestal* XII(1/2): 45-50.

Hoganson, H.M. and Borges, J.G. 1998. Using dynamic programming and overlapping subproblems to address adjacency in large harvest scheduling problems. *Forest Science* 44: 526-538.

Hoganson, H.M. and Borges, J.G. 2000. Impacts of the time horizon for adjacency constraints in harvest scheduling. *Forest Science* 46: 176-187.

Hoganson, H.M., Borges, J.G. and Bradley, D. 1998. Techniques for addressing spatial detail in forest planning. In: P. Bachmann, M. Kohl and R. Päivinen (Eds.), *Assessment of Biodiversity for Improved Forest Planning*. EFI Proceedings no.18. Kluwer Academic Publishers Forestry Sciences Vol. 51: 181-188.

Hoganson, H.M. and Kapple, D. 1991. DTRAN 1.0: *A multi-market timber supply model*. College of Natural Resources and Agricultural Experiment Station, Department of Forest Resources Staff Paper Series Report No. 82, University of Minnesota, St. Paul, 66 p.

Hoganson, H.M. and Kapple, D. 1995. *Estimating impacts of extended rotation forestry*. Final research report submitted to the Minnesota Department of Natural Resources, 170 p.

Hoganson, H.M. and McDill, M. 1993. *Relating reforestation investment in northern Minnesota with industry needs, nontimber values and mitigation strategies*. Research report submitted to the Charles K. Blandin Foundation, 232 p.

Hoganson, H.M. and Rose, D.W. 1984. A simulation approach for optimal timber management scheduling. *Forest Science* 30: 220-238.

Hunter, M.L. 1990. *Wildlife, forests and forestry. Principles of managing forests for biological diversity*. Prentice Hall, Inc., New Jersey, 370 p.

ILOG, Inc. 1997. CPLEX Base Systems, Incline Village, Nevada.

Jaakko Poyry Consulting, Inc., 1994. *Generic Environment Impact Statement on Timber Harvesting and Forest Management in Minnesota*. Tarrytown, NY, Jaakko Poyry Consulting, Inc., 813 p.

Johnson, K.N., and Tedder, P.L. 1983. Linear programming vs. binary search in periodic harvest level calculation. *Forest Science* 29: 569-581.

Johnson, K.N. and Scheurman, E. 1977. Techniques for prescribing optimal timber harvest and investment under different objectives – Discussion and Synthesis. *Forest Science Monograph* 18, 31 p.

Jonsson, B., Jacobsson, J. and Kallur, H. 1993. The forest management planning package. Theory and application. *Studia Forestalia Suecica* No. 189, 57 p.

Kangas, J. 1999. The Analytic Hierarchic Process (AHP): standard version, forestry applications and advances. In: Helles, F., Anderson, P. and Wichmann, L. (Eds.). 1999. *Multiple-use of forests and other natural resources. Aspects of theory and application*. Kluwer Academic Publishers, Copenhagen, Forestry Sciences 61: 96-105.

Kangas, J., Loikkanen, T., Pukkala, T., and Pykäläinen, J. 1996. A participatory approach to tactical forest planning. *Acta Forestalia Fennica* 251. 24 p.

Kangas, J. and Pukkala, T. 1996. Operationalization of biological diversity as a decision objective in tactical forest planning. *Canadian Journal of Forest Research* 26: 103-111.

Kurttila, M. 2001. *Methods for integrating ecological objectives into landscape-level planning of non-industrial private forestry*. Doctor Academic Dissertation, Faculty of Forestry, University of Joensuu, Joensuu. 35 p.

Lockwood, C., and Moore, T. 1993. Harvest scheduling with spatial constraints: a simulated annealing approach. *Canadian Journal of Research* 23: 468-478.

McDill, M. and Braze, J. 2000. Comparing adjacency constraint formulations for randomly generated forest planning problems with four age-class distributions. *Forest Science* 46: 423-436.

McDill, M. and Braze, J. 2001. Using the branch and bound algorithm to solve forest planning with adjacency constraints. *Forest Science* 47: 403-418.

Metropolis, N., Rosenbluth, A., Rosenbluth, M., Teller, A., and Teller, E. 1953. Equation of state calculations by fast computing machines. *Journal of Chemical Physics* 21: 1087-1091.

Michalewicz, Z. 1996. *Genetic algorithms + data structures = evolution programs*, 3rd ed., Springer-Verlag, Berlin, 387 p.

Murray, A. 1999. Spatial restrictions in harvest scheduling. *Forest Science* 45: 45-52.

Murray, A. and Church, R. 1995. Heuristic solution approaches to operational forest planning problems. *OR [Oper. Res.] Spektrum* 17: 193-203

Murray, A. and Church, R. 1996. Constructing and selecting adjacency constraints. *Infor* 34: 232-247.

Navon, D.I. 1987. An evaluation of FORPLAN from an operations research perspective: discussion paper. In: *Proceedings of FORPLAN symposium*, USDA Gen. Tech. Rep. RM-140 pp.147-154.

Nelson, J.D. and Brodie, J.D. 1990. Comparison of a random search algorithm and mixed integer programming for solving area-based forest plans. *Canadian Journal of Research* 20: 934-942.

Nelson, J.D., Brodie, J.D. and Sessions, J. 1991. Integrating short-term, area-based logging plans with long-term harvest schedules. *Forest Science* 37: 101-122.

O'Hara, A.J., Faaland, B.H. and Bare, B.B. 1989. Spatially constrained timber harvest scheduling. *Canadian Journal of Research* 19: 715-724.

Öhman, K. 2000. Creating contiguous areas of old forest in long-term forest planning. *Canadian Journal of Research* 30: 1817-1823.

Öhman, K. and Erikkson, L.O. 1998. The core area concept in forming contiguous areas for long term forest planning. *Canadian Journal of Research* 28: 1032-1039.

Paredes, G. and Brodie, J.D. 1988. Activity analysis in forest planning. *Forest Science* 34: 3-18

Paredes, G. and Brodie, J.D. 1989. Land value and the linkage between stand and forest-level analysis. *Land Economics* 65(2): 158-166.

Pham, D.T. and Karaboga, D. 2000. *Intelligent optimisation techniques. Genetic algorithms, tabu search, simulated annealing and neural networks.* Springer-Verlag, London, 302 p.

Press, W.H., Teukolsky, S.A., Vetterling W.T. and Flannery, B.P. 1992. *Numerical Recipes in C.* Cambridge University Press. 994 p.

Pukkala, T. and Kangas, J. 1993. A heuristic optimization method for forest planning and decision-making. *Scandinavian Journal of Forest Research* 8: 560-570.

Pukkala, T. and Kangas, J. 1996. A method for incorporating risk and risk attitude into forest planning. *Forest Science* 42: 198-205.

Pukkala, T., Nuutinen, T. and Kangas, J. 1995. Integrating scenic and recreational amenities into numerical forest planning. *Landscape and Urban Planning* 32: 185-195.

Pykäläinen, J. 2000. *Interactive Use of Multi-Criteria Decision Analysis in Forest Planning*. Doctor Academic Dissertation, Faculty of Forestry, University of Joensuu, Joensuu. 37 p.

Reeves, C.R. (ed.) 1993. *Modern heuristic techniques for combinatorial problems.* Ed 1. John Wiley & Sons, Inc., New York, 320 p.

Richards, E., and Gunn, E. 2000. A model and tabu search method to optimize stand harvest and road construction schedules. *Forest Science* 46: 188-203.

Rose, D.W. 1984. A critical review of mathematical programming in forest resource planning. In: *Proceedings of the IUFRO Symposium on Forest Management Planning and Managerial Economics*, Tokyo, pp. 306-317.

Rose, D.W. 1999. Data, models and technology in forest planning. *Revista Florestal* XII(1/2): 35-44.

Rose, D.W. and Husain, S.A. 1998. Shadow pricing trade-offs between timber supply and environmental quality. In: P. Bachmann, M. Köhl and R. Päivinen (Eds.), *Assessment of Biodiversity for Improved Forest Planning*. EFI Proceedings no.18. Kluwer Academic Publishers Forestry Sciences Vol. 51: 111-119.

Saaty, R. 1977. A scaling method for priorities in hierarchical structures. *Journal of Mathematical Psychology* 15: 234-281.

Saaty, R. 1980. *The Analytic Hierarchic Process. Planning, priority setting, resource allocation.* McGraw-Hill Publishing Company, New York, 283 p.

Sessions, J. 1992. Solving for habitat connections as a Steiner network problem. *Forest Science* 38: 203-207.

Snyder, S. and ReVelle, C. 1997. Dynamic selection of harvests with adjacency restrictions: the SHARe model. *Forest Science* 43: 213-222.

Strange, N., Tarp. P., Helles, F. and Brodie, J. 1999. A four-stage approach to evaluate management alternatives in multiple-use forestry. *Forest Ecology and Management* 124: 79-91.

Tarp, P. and Helles, F. 1997. Spatial optimization by simulated annealing and linear programming. *Scandinavian Journal of Forest Research* 12: 390-402

Van Deusen, P. 1996. Habitat and harvest scheduling using bayesian statistical concepts. *Canadian Journal of Forest Research* 26: 1375-1383.

Van Deusen, P. 1999. Multiple solution harvest scheduling. *Silva Fennica* 33: 207-216.

Van Deusen, P. 2001. Scheduling Spatial Arrangement and harvest simultaneously. *Silva Fennica* 35: 85-92.

Weintraub, A., Barahona, B. and Epstein, R. 1994. A column generation algorithm for solving general forest planning problems with adjacency constraints. *Forest Science* 40: 142-161.

Weintraub, A., Church, R., Murray, A. and Guignard, M. 2000. Forest management models and combinatorial algorithms: analysis of the state of the art. *Annals of Operations Research* 96: 271-285.

Williams, H.P. 1990. *Model building in mathematical programming*, 3rd ed., John Wiley & Sons, Chichester, 356 p.

Zeleny, S.H. and Evans, J.R, 1981. Heuristic "optimization": why, when and how to use it. *Interfaces* 11(5): 84-91.

Chapter 7

SPATIAL OPTIMISATION IN FOREST PLANNING

A review of recent Swedish research

KARIN ÖHMAN

Department of Forest Resource Management and Geomatics, SLU, SE-901 83 Umeå, Sweden

Abstract. There is an increasing interest in considering goals other than timber production in forest planning. Common to many of these new goals is that there are spatial requirements. Unfortunately, existing approaches for long-term forest planning are in general non-spatial, or are not effective in spatial considerations because of the complexity of solving spatial problems. These circumstances have spurred extensive research into spatial problem solving during the last years. However, much of the research has a focus on American conditions. Subsequently, a majority of the studies deal with the problem of not harvesting adjacent areas and only a small number of studies deal with problems more typical to, e.g., Scandinavian conditions, such as minimizing the fragmentation of old forest. In this chapter different aspects of spatial consideration will be discussed. Different ways of including spatial aspects in forest planning, and characteristics of spatial problems, will be presented. Further, a number of planning problems typical for a Scandinavian setting that encompass both spatial considerations in terms of creating connectivity in the landscape and also non-spatial considerations that normally are found in long term forest planning, will be presented.

Keywords: connectivity problems, harvest scheduling, landscape structure, linear programming, simulated annealing.

Timo Pukkala (ed.), Multi-objective forest planning, 153–172.

1. INTRODUCTION

Over the last few years there has been increased interest in optimising the spatial layout of forest management actions. Unfortunately, the existing approaches to forest planning to a large extent lack the means to effectively include spatial considerations in numerical optimisation. Further, the focus in many suggested approaches has been on how to include consideration to adjacency or green-up constraints in forest planning (restrictions that prevent harvest in adjacent areas for a certain time period). However, in maintaining forest biodiversity, methods that approach, for instance, the problem with the fragmentation of old forest also have to be developed. If the goal is to minimize fragmentation it could be more effective to establish areas of undisturbed forest rather that reduce the cutting intensity across the entire landscape (Franklin and Forman 1987, Gustafson and Crow 1996). It has been shown that clustering harvests in time and space would allow more forest interior habitats to be sustained (Gustafson 1996). Further, when maintaining species that need forest interior conditions, it could be better to maximize the core area (the undisturbed forest free of edge effect) rather than maximizing the total old forest area (Fig. 1) (Temple 1985). However, spatial evaluation of different management options is necessary because for financial as well as ecological reasons. For example, it is generally recognized that it is more expensive to harvest widely dispersed areas than areas that are clustered in space (Baskent and Jordan 1991, Gustafsson 2000).

Long-term forest management that considers spatial features consists of managing different area elements in such a way that a good spatial pattern will arise while at the same time economic goals are satisfied. Traditionally, considerations of spatial relationships are not included in planning at the strategic level. Instead, harvest and silvicultural activities are scheduled in the tactical planning stage to meet the defined spatial constraints. Unfortunately, omitting the spatial constraints and objectives at the strategic level may lead to misleading results, or plans that are impossible to implement (Daust and Nelson 1993, Clements et al. 1990). Instead of including spatial considerations at the strategic level it is often merely required that a certain amount of old forest (for instance) be retained. But using aggregate amounts of habitats at the strategic level, when the patch shape, size and distance between habitats patches are all important, can result in overestimates of cutting possibilities or overestimates of the amount of habitat that the species really can use (Davis et al. 2000).

In this chapter, aspects of spatial consideration in forest planning will be discussed. First, various ways of including spatial aspects in forest planning, and characteristics of spatial problems, will be presented. Second, focus will shift to planning problems that encompass both spatial considerations in terms of creating connectivity in the landscape and considerations typical to traditional long term forest planning in Scandinavia.

2. SPATIAL CONSIDERATIONS IN THE OPTIMIZATION PROCESS

2.1. Exogenous and endogenous approaches to spatial considerations

In general there are two main approaches, the exogenous and the endogenous approach, for incorporating spatial considerations into the planning process (Kurttila 2001). The exogenous approach is a two-step procedure in which the optimisation does not include any spatial information but takes into account predetermined constraints derived from spatial considerations. This can be done by manipulating the simulation of permitted treatment schedules in such a way that only schedules that generate a certain spatial structure are allowed (e.g., Nalli et al. 1996, Naesset 1997, Fries and Lämås 2000, Kangas et al. 2000). An example of an exogenous approach to spatial planning is from ecological landscape planning (ELP), a recently developed tool for balancing timber production with the maintenance of biodiversity at the landscape level (Törnquist 1996). In ELP, areas set aside are first distributed as management zones; for instance, buffers are preserved around streams and lakes or key biotopes are set aside. Often, this dellmitation is made subjectively. Second, the long-term planning problem is solved within the fixed set of zones without explicit reference to spatial information.

In the endogenous approach, the spatial part is solved together with other problem elements in one step, i.e., variables that describe the spatial relationships are processed by the solution algorithm, and the spatial layout is generated in the optimisation together with the economic objectives (Kurttila 2001). Examples of variables that describe spatial relationships between stands and that can be included in the optimisation are, e.g., amount of core area, common boundaries between stands, contiguous clear cut area, etc. (e.g., Lockwood and Moore 1993, Kerkervorde 1996, Öhman and Eriksson 1998). With the endogenous approach a huge number of different spatial layouts can be evaluated. This is particularly important in long-term planning where the efficient enhancement of biodiversity requires that plans are adapted to the future development of the forest, i.e., zones are not delimited once and for all, but could change over time. Thus, the endogenous approach has potential to generate plans that are more effective from both the economic and ecological point of view.

2.2. Characteristics of spatial problems

Existing systems for long-term forest planning are in general non-spatial, e.g., FMPP in Sweden (Jonsson et al. 1993), MELA in Finland (Siitonen and Nuutinen 1996), FORPLAN in USA (Johnson et al. 1986) and FOLPI in New Zealand (Manley et al. 1991). This necessitates an exogenous approach to planning, in which zones are delimited before using the optimisation routines of the systems. If, instead, consideration of spatial relationships is included in an endogenous way in long term planning, the complexity of the task would increase (Hof and Bevers

1998, Krcmar-Nozic et al. 1998, Martell et al. 1998) for a number of reasons. First, it is no longer possible to use strata-based planning in, e.g., FMPP. Instead, to realistically model spatial relationships between individual units, location-specific data are needed (Daust and Nelson 1993, Church et al. 1998). Second, to represent spatial relationships, integer variables must be used (Daust and Nelson 1993, Murray and Snyder 2000). The next problem concerns accounting for the output response to different spatial configurations of management actions. Typically, in non-spatial forest planning the output response, e.g., the volume harvested in a certain period, is expressed as the sum over a number of treatment schedules for a number of management units. However, problems that involve spatial consideration are often affected by interdependencies between individual management units, i.e., the output response is dependent on actions taken in neighbouring units (Hof and Bevers 1998, Hoganson et al. 1998). More precisely, the production functions for the outputs of the units are no longer homogeneous of degree one with respect to area. Thus, the outputs can no longer be expressed in terms of a linear combination of treatment schedules (Eriksson 1983). This causes difficulties, since combinations of non-linear functions and integer variables are difficult to solve and many of the planning systems available today are designed to handle linear functions and continuous variables, e.g., FOLPI (Manley et al. 1991), FORPLAN (Johnson et al. 1986), GAYA-LP (Hoen 1996), MELA (Siitonen and Nuutinen 1996), and SPECTRUM (Camenson et al. 1996).

3. SPATIAL PROBLEMS

3.1. The adjacency problem

The difficulties mentioned above, and the increased knowledge about spatial aspects, have spurred extensive research into spatial problem solving. However, much of the research is focused on avoiding large open clear-cuts in the landscape (e.g., Torres-Rojo and Brodie 1990, Nelson and Finn 1991, Dahlin and Sallnäs 1993, Snyder and Revelle 1996a, Murray 1999, McDill and Braze 2000), but only a few studies deal with problems more typical to Scandinavian conditions, such as the fragmentation of old forest (Öhman 2000) or the design of corridors between habitats (Sessions 1992, Williams 1998).

The adjacency problem is characterized by a restriction that a contiguous clear-cut area should not exceed a certain size. Further, before an area adjacent to a clear-cut can be cut, a certain time must pass. An unfortunate consequence of adjacency constraints is increasing fragmentation, since the cuttings are dispersed over the landscape (Barrett et al. 1998). Two different approaches have been used to control the size of clear-cuts: the unit restriction model (URM) and the area restriction model (ARM) (Murray 1999). In the URM, the sizes of individual management units are assumed to be near the maximum opening size. If one unit is scheduled for harvest, all adjacent units are restricted from harvest until the regeneration in the cut unit has progressed for a defined minimum time. The URM can be formulated as either an integer programming (IP) problem or a mixed integer programming (MIP) problem, and exact techniques can be used to solve the problem (Snyder and ReVelle 1997). However, in many cases the URM has been solved with heuristic methods because of the large size of the problem. In mathematical programming, the traditional approach is to formulate one constraint for each pair of neighbouring management units (e.g., Nelson and Brodie 1990, Murray and Church 1995, Snyder and ReVelle 1996b). The general form of the pairwise constraint for a one-period problem is:

$$X_i + X_j \leq 1 \qquad \forall i \in I, \forall j \in N^i \qquad (1)$$

where i is the index, I is the set of planning units, and N^i is the set of units adjacent to i. If unit i is to be harvested then X_i=1 and if unit i is not to be harvested then X_i=0. Since the pairwise approach generates a large number of constraints, many studies have aimed to decrease the number of necessary constraints (Jones et al. 1991, Murray and Church 1995, McDill and Braze 2000). An example of an approach that reduces the number of constraints is the compartmental approach, in which one constraint is generated for each management unit and all the units adjacent to it (Torres-Rojo and Brodie 1990, Murray and Church 1995). The general form of the compartmental constraint for a one period problem is:

$$n_i X_i + \sum_{j \in N^i} X_j \leq n_i \qquad \forall i \in I \qquad (2)$$

where n_i is the number of stands that are adjacent to stand i. However, the number of constraints is less important today as much of the software accepts formulations with almost an unlimited number of constraints. Therefore much of the work being done today instead concentrates on reducing the solution time and the difficulties involved in generating the constraints (McDill and Braze 2000, Weintraub et al. 2000).

In the ARM it is assumed that the sizes of individual treatment units are well below the maximum opening size. Harvesting adjacent units is therefore allowed as long as the total contiguous harvested area is less than the maximum opening size. ARMs are much more difficult to solve than URMs, since it is not possible to decide in advance all the possible combinations of harvested units (Murray 1999, Richards and Gunn 2000). The possibilities of solving ARMs with exact methods are therefore limited. These problems have instead been solved with heuristic methods because of the non-linear characteristics of the area constraint (Lockwood and Moore 1993, Clark et al. 2000, Richards and Gunn 2000).

3.2. Connectivity problems

Another type of spatial problem is one that focuses on creating connectivity in the landscape. While adjacency problems focus on keeping spatial elements with certain conditions, e.g., clear-cuts apart from each other in the landscape, connectivity problems try to aggregate stands with certain conditions together. A typical example of a connectivity problem is to create contiguous areas of old forest in order to minimize fragmentation. Unfortunately, there have been very few studies that aim at aggregating the old forests in the optimisation component of long term forest planning. A few examples of problems that result in aggregated areas of old forest can be found, e.g., in Hof and Joyce (1993), Clements et al. (1999) and Nalle et al. (2002). Further, studies that aim at aggregating resources often include a limited number of spatial elements, consisting of only a single time period or do not include an economic evaluation of different options. However, for several realistic spatial problems, a larger set of spatial elements or a longer planning horizon is needed to properly reflect the nature of the problem.

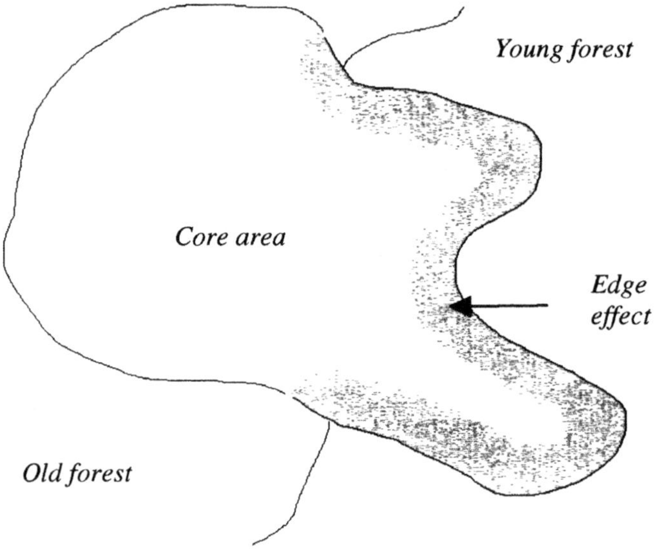

Figure 1. The core area of a stand is the area that is not influenced by edge effects from surrounding stands. The stand in the figure is surrounded partly by young forest that causes edge effects and partly by old forest that does not cause edge effects. As a result only a part of the stand consists of core area.

The fragmentation problem

An example of a problem that needs a larger set of spatial elements is the formation of core area in order to minimize the fragmentation of old forest. Core area is defined as the area of a stand that is not influenced by edge effects from surrounding habitats and is a function of patch size, shape and nature of immediately surrounding conditions (Fig. 1) (Zipper 1993, Baskent and Jordan 1995).

The core area concept was tested in two studies by Öhman and Eriksson (1998) and Öhman (2000) as a criterion for promoting the formation of contiguous old forest in the landscape over time. The concept was evaluated in both studies by solving a long term planning problem consisting of maximizing the economic efficiency while creating contiguous areas of old forest. In Öhman and Eriksson (1998) the planning problem consisted of maximizing the net present value (NPV) over an infinite time horizon subject to a certain demand for core area in each period. The purpose of using the core area concept in the studies was to express whether stands consisting of old forest were adjacent. In keeping with the definition of core area, the amount of core area will increase if adjacent stands of old forest are saved. The stated problem was solved with simulated annealing (SA) (Laarhoven and Aarts 1987). As a result the core area demands were accounted for by penalty functions added to the objective function. The mathematical formulation of the problem in Öhman and Eriksson (1998) was as follows:

Maximise

$$z = \sum_{i=1}^{I}\sum_{j=1}^{J_i}\left(D_{ij} + \overline{D}_{ij}\cdot Y_i\right)\cdot X_{ij} - \alpha\cdot\sum_{p=1}^{P}\left(\overline{C}_p - \sum_{i=1}^{I} C_{ip}\right)^2_{|>0} \tag{3}$$

subject to

$$C_{ip} = \Omega_{ip}(\mathbf{X}) \qquad i = 1,\ldots,I;\ p = 1,\ldots P \tag{4}$$

$$C_{iP}\cdot Y_i = 0 \qquad i = 1,\ldots I \tag{5}$$

$$\sum_{j=1}^{J_i} X_{ij} = 1 \qquad i = 1,\ldots I \tag{6}$$

$$X, Y = \{0,1\} \tag{7}$$

Where I = total number of stands; J_i= total number of treatment schedules for stand i; D_{ij} = NPV from period 1 to period P from stand i and treatment schedule j; $\overline{D}_{ij}$ = NPV from P+1 to infinity from stand I and treatment schedule j; X_{ij} = binary decision variable that takes the value of 1 if stand i is assigned to treatment schedule j; P = total number of periods with requirements for the spatial structure; Y_i = dummy variable that is forced to take on the value 0 if a stand has core area in period P, i.e., if $C_{iP} > 0$; $\overline{C}_p$ = core area desired in period p; C_{ip} = core area in stand i in period p; $\Omega_{ip}(\mathbf{X})$ = scalar valued operator on $\mathbf{X} = \{X_{11},\ldots,X_{IJ_I}\}$ that yields the core area of stand i in period p; α = penalty parameter associated with the requirement for core area.

The first product sum in Equation 3 gives the NPV and the second term gives the penalty due to non-fulfilment of the core area requirement. Equation 4 gives the core area of each stand in each period. Given that a stand has core area in period P,

Equation 5 together with Equation 7 ensures that the stand is treated as an area left for free development after that period, and the NPV after period P is not added to the objective function value. Equation 6, together with Equation 7, ensures that a stand is allocated only one schedule.

The model formulation was supplemented in Öhman (2000) with a criterion concerning the amount of edge habitats, requiring that the difference between the total amount of old forest and the amount of core area should be minimzed. The purpose of this criterion was to allow additional weight to be placed on aggregation. Also in this study the problem was solved with SA and the spatial requirements were accounted for by penalty functions. The mathematical formulation of the problem solved in Öhman (2000) was as follows:

Maximise

$$z = \sum_{i=1}^{I} \sum_{j=1}^{J_i} (D_{ij} + \overline{D}_{ij} \cdot Y_i) \cdot X_{ij} - \alpha \cdot \sum_{p=1}^{P} \left(\overline{C}_p - \sum_{i=1}^{I} C_{ip} \right)^2_{|>0} - \beta \cdot \sum_{p=1}^{P} \left(\sum_{i=1}^{I} G_{ip} - \sum_{i=1}^{I} C_{ip} \right)^2 \quad (8)$$

subject to Equations 4–7 and to:

$$G_{ip} = \sum_{j=1}^{J_i} G_{ijp} X_{ij} \qquad i = 1,...I; p = 1,...P \quad (9)$$

where G_{ijp} = area of old forest in stand i for treatment schedule j in period p; β = the weight associated with the amount of edge habitat.

The model formulation was applied in Öhman and Eriksson (1998) to a simulated landscape consisting of 200 stands while in Öhman (2000) it was applied to a real landscape consisting of 755 stands. The spatial requirements were valid in both studies during the first 100 years. In Öhman and Eriksson (1998) the definition of old forest was based on an age criterion in which two different ages, 80 and 120 years, were tested. Furthermore, only stands younger than 50 years were considered to cause edge effects on surrounding old forest habitats. Finally, two different edge widths for calculating the core area, 32 and 64 meters, were tested. The definition of old forest used in Öhman (2000) was linked both to an age criterion and to previous thinnings. To simplify calculations, all stands that were not composed of old forest, wetlands, impediments or lakes were assumed to cause edge effects on surrounding habitats. As in Öhman (2000), two different edge widths were tested: 30 and 60 m.

The model formulations in both Öhman and Eriksson (1998) and Öhman (2000) created contiguous areas of old forest over time in the landscape (Fig. 2). However, the results from Öhman (2000) are to some extent in contrast to the results in Öhman and Eriksson (1998). In Öhman (2000) the requirement for a certain amount of core area did not alone create aggregations of old forest. Further, in Öhman (2000) an increase in the edge effect caused only a marginal increase in the aggregation of old forest, which also is in contrast to Öhman and Eriksson (1998) where an increase in edge effect caused an increase in the aggregation of old forest.

The cost of taking spatial considerations into account was modest in Öhman and Eriksson (1998). However, in Öhman (2000) the decrease in NPV was significant when both the core area requirement and the amount of edge habitats were included in the model formulation.

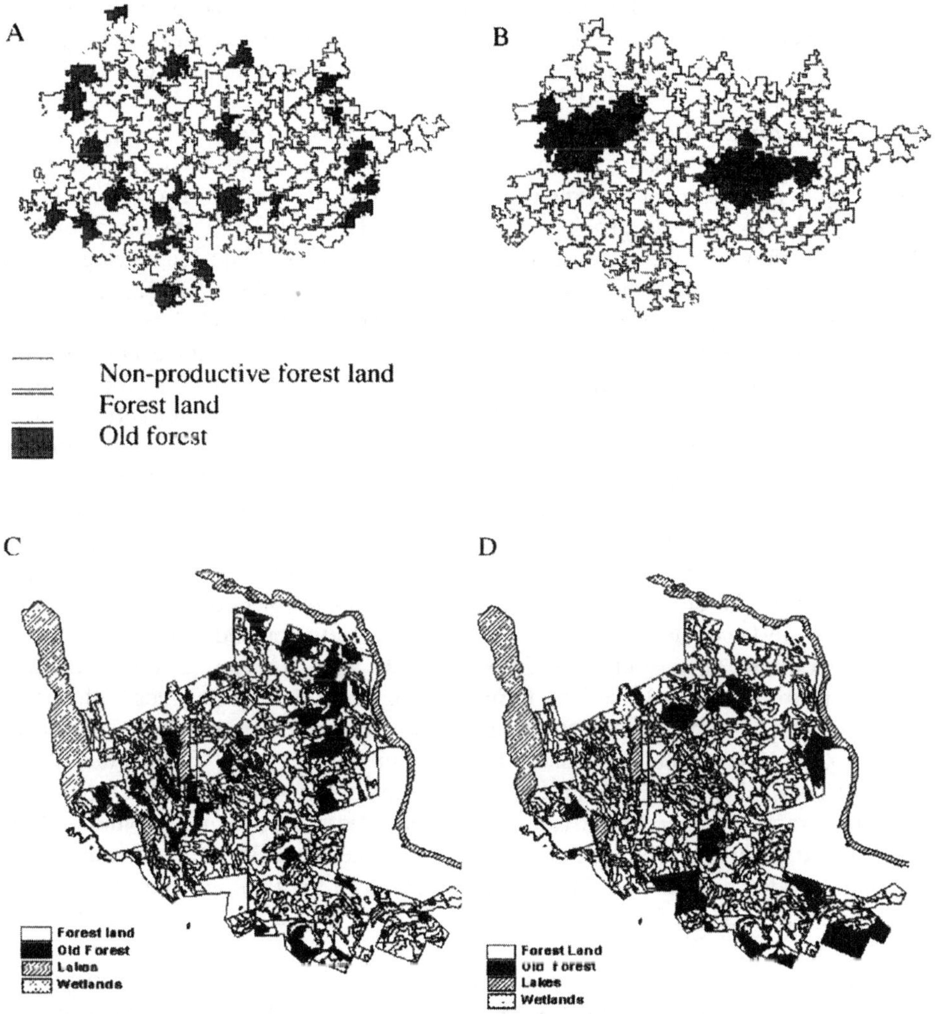

Figure 2. The formation of old forest after 100 years a) without consideration to the amount of core area (Öhman and Eriksson 1998) b) with consideration to the amount of core area (Öhman and Eriksson 1998) c) without consideration of the amount of edge habitats (Öhman 2000) d) with consideration of the amount of edge habitats (Öhman 2000)

One explanation for the difference in results between the two papers is that it was more difficult to create old forest in Öhman (2000) because of a more complicated definition of old forest. In Öhman and Eriksson (1998), stands forming a contiguous area of old forest harvested at the same time became old forest in the same time frame in the future. This is different from Öhman (2000), in which two stands that were harvested at the same time did not necessarily become old forest in the same period. Further, in Öhman (2000) there was an extra cost associated with creating old forest since profitable thinnings may have had to be omitted.

The general forest planning problem with spatial considerations

One limitation with many spatial studies is that they deal with small sample problems that only aim at maximizing, e.g., the NPV, while creating a certain spatial structure. Most of these problems have been solved with MIP or some kind of heuristic method. However, even if these methods are suitable for solving spatial problems they have disadvantages such as the computational burden that is invoked, in order to find good solutions for large problems. Practical forestry requires methods that can be applied to large landscapes consisting of thousands of units where other forest wide restrictions are also included in the problem formulation. Linear programming (LP) is one of the most widely used methods for solving forest planning problems (Johnson and Scheurman 1977). One reason for the popularity of LP is its computational efficiency. An LP model can govern almost any quantity that can be formed as a sum over a number of units and treatment schedules. However, because of the linearity assumptions of LP, it is difficult to handle spatial problems with this method. In summary, both LP and heuristics have their pros and cons. It would therefore be favourable if the advantages of both methods could be used (e.g., Kerkervorde 1996, Tarp and Helles 1997, Boston and Bettinger 2001).

An example of a study that integrates LP with SA into a single solution procedure for solving a large long term planning problem that includes both spatial and non spatial forest-wide restrictions is Öhman and Eriksson (2002). The non-spatial part of the problem consisted of maximizing the NPV from future forest management subject to the requirements of even flow in harvest and a certain final growing stock volume. These forest-wide constraints are normally found in long term forest planning. The spatial part of the problem consisted of minimizing the amount of edge habitats, i.e., the difference between the total amount of old forest and the total amount of core area, as well as creating a certain amount of core area each period.

The study tested three different approaches (Fig. 3). In the first approach, the problem was first solved with SA without consideration to the non-spatial forest-wide constraints. The objective function used in the SA algorithm therefore consisted of the NPVs and the penalty functions for the spatial objectives and constraints. In the next step the non-spatial part of the problem was solved with LP. The information that passed from the SA algorithm to the LP part was the set of treatment schedules that gave rise to the same spatial layout of old forest in all periods as given by the solution of the SA algorithm.

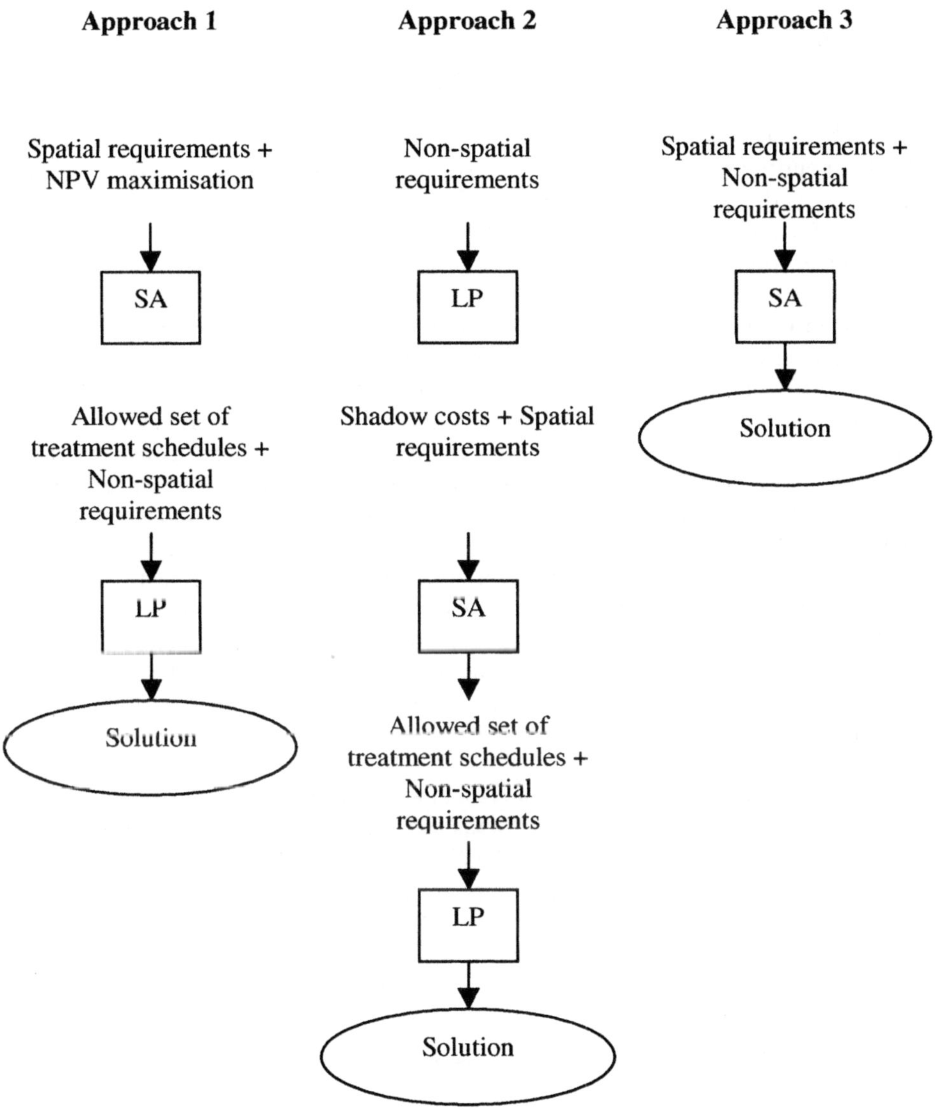

Figure 3. The solution processes for all three approaches (Öhman and Eriksson 2002)

The rationale of the second approach was that the SA algorithm would produce better solutions if it were fed with information about constraint costs derived using LP. Therefore, the second approach began by solving the non-spatial part of the problem with LP, i.e., the NPV maximization problem with consideration of the forest-wide constraints but without consideration of the spatial part of the problem. The solution algorithm then continued with SA for solving the spatial part of the problem. The shadow costs, or the dual solution of the LP problem, were then brought to the SA analysis via the reduced costs of the treatment schedules, i.e., the objective function that was maximized in the SA algorithm consisted of the reduced cost for the treatment schedules together with the penalty functions for the spatial constraints. After the SA algorithm, the solution procedure continued with LP for solving the non-spatial constraints, as in approach 1. In the third approach SA solved the entire problem, i.e., the objective function consisted of the NPV for the treatment schedules together with penalty functions for both the spatial constraints and the non-spatial constraints.

All three approaches were tested in a case study on the same landscape as in Öhman (2000) with the same definitions on old forest, planning horizon etc. The results from the case study indicate that the old forests were less aggregated after the problem was solved with approach 3 than with approaches 1 and 2. On the other hand, the differences in the degree of aggregation between approaches 1 and 2 were very small. Also in terms of NPVs, approach 3 produced worse solutions compared to approach 1 and 2 (Table 1). However, the cost of taking spatial considerations into account was significant for all three approaches as compared to a reference problem (Table 1). The reference problem had the same formulation as the LP problem in approach 1 and 2 except that a minimum requirement for old forest was added to the formulation. In summary, it seems that when more constraints were added, i.e., the entire problem was solved with SA, the algorithm was more inclined to get stuck at local optima. Better solutions could probably have been found with better parameter settings. However, an important issue that should be considered when evaluating any solution method is the problem of finding appropriate parameter settings.

Table 1. NPVs for all approaches and cases (10^6 Swedish crowns). In case 1 the harvest was allowed to fluctuate up and down by 10 % from period to period. In case 2 the harvest had to increase by at least 10 % from period to period.

Approach	*Case 1*	*Case 2*
1	79.3	68.2
2	79.6	68.3
3	72.5	56.5
Reference problem	89.1	

The harvest clustering problem

Another example of a typical connectivity problem is the clustering of harvests in time and space. This could be motivated by savings in entry costs and costs associated with road building (Baskent and Jordan 1991, Gustafsson 2000). In addition, if the harvests are clustered, more interior forests are preserved in the landscape (Gustafson 1996). Normally in forest planning models the NPV only includes the discounted revenues from timber sales, and costs of harvest operations and regeneration from period one to infinity. The costs do not include entry (transportation of harvesters, road maintenance, etc.) or timber transportation. As a result, NPV maximization alone will not lead to a clustering of harvest activities. However, including the entry cost in the NPV calculations is difficult since defining all potential contiguous harvest areas in advance is infeasible even for small problems. Another way to approach the problem is to include other restrictions in the model formulation that lead to clustering of harvest activities.

Öhman and Lämås (2001) presented a model for delineating harvest areas in the landscape. Two objectives were addressed in the model. The first was to maximize the NPV of future forest management and the second was to cluster the harvests areas in time and space. In the model, the approach for clustering the thinnings and final cuttings was based on a criterion denoted effective volume (EV). The EV should be seen as a spatial modelling concept for indicating whether harvests take place in adjacent stands and periods, not as a measurement of the volume harvested. The two-objective problem was converted to a single objective problem by weighting the two objectives together. The mathematical formulation of the problem was as follows:

Maximise

$$z = w_1 \sum_{i=1}^{I} \sum_{j=1}^{J_i} D_{ij} X_{ij} + w_2 \sum_{i=1}^{I} \sum_{p=1}^{P} EV_{ip} \tag{10}$$

subject to:

$$EV_{ip} = \sum_{j=1}^{J_i} Y_{ijp} X_{ij} \sum_{l}^{N^i} \sum_{j}^{J_l} \sum_{t=1}^{P} V_{ljt} X_{lj} e^{-\beta|p-t|k} \qquad \forall i \in I, \forall p \in P \tag{11}$$

$$\sum_{i=1}^{I} \sum_{j=1}^{J_i} V_{ijp} X_{ij} = \overline{V}_p \qquad \forall p \in P \tag{12}$$

$$\sum_{j=1}^{J_i} X_{ij} = 1 \qquad \forall i \in I \tag{13}$$

$$X_{ij} = \{0,1\} \qquad \forall i \in I, \forall j \in J \tag{14}$$

$$w_1 + w_2 = 1 \tag{15}$$

$$w_1, w_2 \in [0,1] \tag{16}$$

where I = total number of stands; J_i= total number of treatment schedules for stand i; P = total number of periods; k = period length; D_{ij} = the NPV for stand i and treatment schedule j; EV_{ip} = the effective volume for stand i in period p; X_{ij} = binary variable that indicates whether stand i is assigned to treatment schedule j; N^i = the set of stands that are adjacent to stand i; including stand i itself; Y_{ijp}= the coefficient for indicating that treatment schedule j for stand i has a thinning or clear cutting in period p; V_{ijp} = volume harvested from stand i, under treatment schedule j, in period

$p; \overline{V}_p$ = the volume demand in period p; β = the factor that influences the effect of harvest in adjacent periods on the amount of EV; w_1 = the weight for the first objective, the NPV maximization; w_2 = the weight for the second objective, the EV maximization.

Equation 10 gives the weighted sum of NPV and EV. Equation 11 expresses the EV for stand i in period p. EV_{ip} is the sum of harvested volumes from stand i and neighbouring stands in period p and adjacent periods t. For each p, t goes from 1 to P. However, the volumes harvested in period t are discounted by $e^{-\beta|p-t|k}$. By setting β to a high value the expression $e^{-\beta|p-t|k}$ will be close to zero for all $t \neq p$ and only harvest activities in the same period will contribute to the amount of EV for adjacent stands. On the other hand, by setting β to a low value, close to zero, it does not matter in which period the harvest takes place for contributing to the amount of EV as long as the harvest activities take place in adjacent stands. Equation 12 ensures that the volume demand is fulfilled in each period. Equation 13 together with Equation 14 ensures that all stands are assigned one treatment schedule. Finally Equations 15 and 16 ensure that the sum of the two weights is one.

The presented model was applied in a case study to a landscape consisting of 2643 stands where substantial consideration was paid to other aspects related to biodiversity and recreation. The planning horizon was divided into 5-year periods, where the clustering requirements extended over the first 40 years. In the case study different values of β were evaluated. Since the spatial dimension in the suggested model formulation leads to a problem that is difficult to solve with ordinary optimization techniques, the ensuing problem was solved with SA.

The results from the case study indicated that the presented model is effective for clustering the harvest activities (Figs. 4 and 5). The difference in numbers of clusters between two solutions with high respective low weight on the EV was significant. Further it is shown that the clustering is more pronounced later in the planning horizon (Fig. 6). The results also indicate that it is possible to aggregate the harvest with only a small sacrifice of NPV. The difference in NPV between the two solutions in Figure 6 is only about 7%. The decrease in NPV could be motivated by the fact that costs not included in the presented model, e.g., entry costs and costs associated with road building will probably decrease when the harvests are aggregated in time and space.

Figure 4.The harvest areas in period 5 corresponding to a solution with low weight on effective volume.

Figure 5. The harvest areas in period 5 corresponding to a solution with high weight on effective volume.

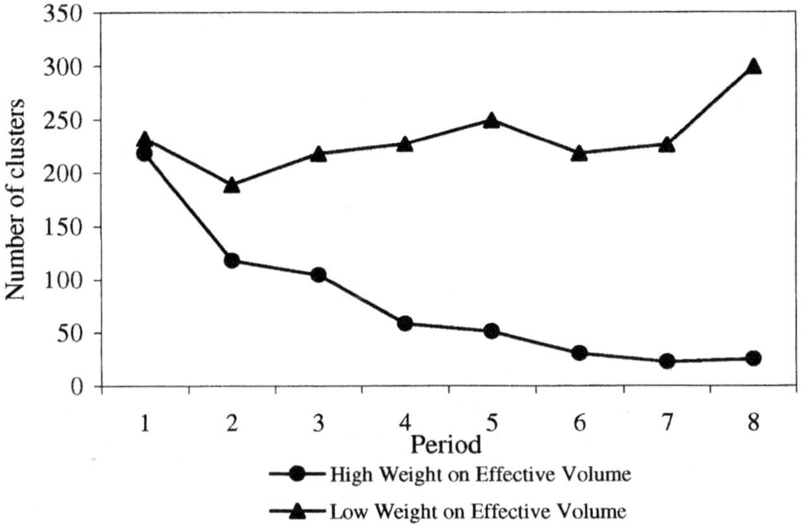

Figure 6. Comparison of the numbers of clusters in each period when the effective volume was given high and low weighting, respectively (Öhman and Lämås 2001). To simplify the calculations in the evaluation a cluster was defined here as a contiguous area of stands harvested in one period. However in the model used in the optimisation, harvests in adjacent periods also contribute to the amount of effective volume.

4. DISCUSSION AND CONCLUSIONS

The interest in taking spatial relationships into consideration has increased. This has spurred an extensive research into solving spatial forest planning problems. However, a majority of the studies deal with the problem of how to avoid harvesting adjacent areas, whereas only a small number deals with problems such as planning management activities while avoiding fragmentation or creating corridors between valuable habitats. This review demonstrated some possibilities for solving forest planning problems that, except for maximizing the NPV of future forest management, also aim at creating connectivity in the landscape, in terms of contiguous areas of either old forest or harvest activities. The motives for including such consideration are both ecological and economic. The emphasis of the demonstrated problems was on reflecting real life situations. Of course there are a number of spatial problems that are not covered by these few examples. Still, these examples could be used for indicating how other spatial planning problems might be approached.

Even if it is not easy to compare the time needed for solving the problem and the cost of spatial considerations between different problems, some general findings can be discussed. First, both the choice of solution method and the criteria affect the solution time. In all four presented studies the spatial parts of the problems were solved with SA. Other methods capable of solving non-linear problems could have been selected, e.g., tabu search or genetic algorithms. However, SA seemed for these studies to be an expedient choice considering the effort of implementing the algorithm and the time needed for solving the problem. Further, the results from Öhman and Eriksson (2001) indicate that one way to shorten the solution times is to combine an exact solution method with a heuristic method. It is also important to select criteria that perform well computationally so that they can be efficiently included in the optimisation model in terms of the time needed for solving the problem (Öhman 2001). The cost of aggregating the harvests seems to be low. This could be explained by the fact that the costs of moving the harvest one period forward or backward are low. On the other hand the costs of aggregating the old forest seem to be more significant. However, the cost is probably highly dependent on, e.g., the definition of old forest and the initial conditions in the landscape.

In conclusion, it appears possible to include considerations of spatial relationships in long-term forest planning even when the problems are of a size found in real-world situations. However, approaches other than traditional methods are often needed. For problems in which the forest-wide constraints are few and only relate to the spatial aspects, it would seem that heuristics alone are adequate. When more forest-wide constraints are added to the problem, a suitable approach could be to combine two solution techniques into one integrated hybrid procedure such as, for example, SA and LP.

REFERENCES

Barret, T.M., Gilless, J.K. and Davis, L.S. 1998. Economic and fragmentation effects of clearcut restrictions. *Forest Science* 44(4): 569-577.

Baskent, E.Z. and Jordan, G.A. 1991. Spatial wood supply simulation modelling. *The Forestry Chronicle* 67(6): 610-621.

Baskent, E.Z. and Jordan G.A. 1995. Characterzing spatial structure of forest landscapes. *Canadian Journal of Forest Research* 25: 1830-1849.

Boston, K. and Bettinger, P. 2001. Development of spatially feasible forest plans: a comparison of two modeling approaches. *Silva Fennica* 35(4): 425-435.

Camenson, D.M., Sleavin, K.E. and Greer, K.E. 1996. SPECTRUM: An analytical tool for building natural resource management models. Large-scale forestry scenario models: experiences and requrements, Joensuu, Finland, 15-22 June, 1995, *EFI proceedings* 5: 133-141.

Church, R. L., Murray, A.T. and Weintraub, A. 1998. Locational issues in forest management. *Location Science* 6: 137-153.

Clark, M.M., Meller, R.D. and McDonald, T.P. 2000. A three-stage heuristic for harvest scheduling with access road network development. *Forest Science* 46(2): 204-218.

Clements, S.E., Dallain, P.L. and Jamnick, M.S. 1990. An operational, spatially constrained harvest scheduling model. *Canadian Journal of Forest Research* 20: 1438-1447

Clements, M.A., ReVelle, C. and Williams, J.C. 1999. Reserve design for species preservation. European *Journal of Operational Research* 112: 273-283.

Dahlin, B. and Sallnäs, O. 1993. Harvest scheduling under adjacency constraints - a case study from the Swedish sub alpine region. *Scandinavian Journal of Forest Research* 8: 281-290.

Daust, K.D. and Nelson, J.D. 1993. Spatial reduction factors for strata-based harvest schedules. *Forest Science* 39(1): 152-165.

Davis, L.S., Jonsson, K.N., Bettinger, P.S. and Howard, T.E. (Eds.) 2000. *Forest management.* McGraw Hill. New York. 804 pp.

Eriksson, L.O. 1983. *Quantitative support by linear programming for decisions on silviculture and harvesting* (thesis), Swedish University of Agricultural Sciences, Department of Operational Efficiency. Working paper nr 248. Garpenberg.

Franklin, J.F. and Forman, R.T. 1987. Creating landscape patterns by forest cutting: Ecological consequenses and principles. *Landscape Ecology* 1(1): 5-18.

Fries, C. and Lämås, T. 2000. Different management regimes in a boreal forest landscape: ecological and economical effects. *Studia Forestalia Suecica* 208: 3-25.

Gustafson, E.J. 1996. Expanding the scale of forest management: allocating timber harvests in time and space. *Forest Ecology and Management* 87: 27-39.

Gustafson, E.J. and Crow, T.R. 1996. Simulating the effects of alternative forest management strategies on landscape structure. *Journal of Environmental Management* 46: 77-94

Gustafsson, K., Larsson, M., Källman, U. and Eriksson, L.O. 2000. *Logistical Determinants in Timber Harvesting Schedule - accounting for geographical aspects in medium term forest planning. Logistics in the Forest Sector.* Helsinki, IUFRO and University in Helsinki.

Hoen, H.F. 1996. Forestry scenario modellingfor economic analysis - experiences using the GAYA- JLP model. Large-scale forestry scenario models: experiences and requirements, Joensuu, Finland, 15-22 June, 1995, *EFI proceedings* 5: 79-88.

Hof, J. and Bevers, M. 1998. *Spatial optimization for managed ecosystems.* Columbia University Press. New York. 258 pp.

Hof, J.G. and Joyce, L.A. 1993. A mixed integer linear programming approach for spatially optimizing wildlife and timber in managed forest ecosystem. *Forest Science* 39(4): 816-834.

Hoganson, H., Borges, J. and Bradley, D. 1998. *Techniques for addressing spatial detail in forest planning. Assessment of biodiversity for improved forest planning.* European Forest Institute Proceedings no.18. Kluwer Academic publishers. Pp. 181-188.

Johnson, K.N. and Scheurman, H.L 1977. Techniques for prescribing optimal timber havest and investment under different objectives. Discussion and synthesis. *Forest Science* 18. 30 pp.

Johnson, K.N., Stuart, T.W. and Crim, S.A. 1986. FORPLAN Version 2: An overview. USDA Forest Service Land Management Planning System Section, Washington DC.

Jones, J.G., Meneghin, B.J. and Kirby, M.W. 1991. Formulating Adjacency constraints in linear optimization models for scheduling projects in tactical programming. Forest Science 37(5): 1283-1297.

Jonsson, B., Jacobsson, J. and Kallur, H.1993. The Forest Management Planning Package. Theory and application. *Studia Forestalia Suecica* 189: 1-56.

Kangas, J., Store, R., Leskinen, P. and Mehtätalo L. 2000. Improving the quality of landscape ecological forest planning by utilising advanced decision-support tools. *Forest Ecology and Management* 132: 157-171.

Kerkervorde, M. 1996. A sequential approach in mathematical programming to include spatial aspects of biodiversity in long range forest planning. Swedish University of Agricultural sciences, Department of Forest Resource Management and Geomatics, working report. 63 pp.

Krcmar-Nozic, E., Kooten G.C.v, Vertinsky, I. and Brumelle, S. 1998. An interactive multiobjective approach to harvest decisions in forest planning. *Scandinavian Journal of Forest Research* 13: 357-369.

Kurttila, M. 2001. The spatial structure of forests in the optimization calculation of forest planning - a landscape ecological perspective. *Forest Ecology and Management* 142: 129-142.

Laarhoven, P.J.M.v. and Aarts, E.H.L. 1987. *Simulated annealing: Theory and applications.* Kluwer Academic Publishers, Dordrecht. 187 pp.

Lockwood, C. and Moore, T. 1993. Harvest scheduling with spatial constraints: a simulated annealing approach. *Canadian Journal of Forest Research* 23: 468-478.

Manley, B., Papps, S., Threadgill, J. and Wakelin, S. 1991. Application of FOLPI A linear programming estate modelling system for forest management planning. Forest Research Institute. FRI Bulletin No. 164. Roturua, New Zealand.

Martell, D.L., Gunn, E.A. and Weintraub, A. 1998. Forest management challenges for operational researchers. *European Journal of Operational Research* 104(1): 1-17.

McDill, M.E. and Braze, J. 2000. Comparing adjacency constraint formulations for randomly generated forest planning problems with four age-class distribution. *Forest Science* 46(3): 423-436.

Murray, A.T. 1999. Spatial restrictions in harvest scheduling. *Forest Science* 45(1): 45-52.

Murray, A.T. and Church, R.L 1995. Measuring the efficacy of adjacency constraint structure in forest planning models. *Canadian Journal of Forest Research* 25: 1416-1424.

Murray, A.T. and Snyder, S. 2000. Spatial modeling in forest management and natural resource planning. *Forest Science* 46(2): 153-156.

Naesset, E. 1997. A spatial decision support system for long-term forest management planning by means of linear programming and a geographical information system. *Scandinavian Journal of Forest Research* 12: 77-88.

Nalle, D.J., Arthur, J.L and Sessions, J. 2002. Designing Compact and Contiguous Reserve Networks with a Hybrid Heuristic Algorithm. *Forest Science* 48(1): 59-68.

Nalli, A., Nuutinen, T., and Päivinen, R. 1996. Site-specific constraints in integrated forest planning. *Scandinavian Journal of Forest Research* 11: 85-96.

Nelson, J. and Brodie, J.D. 1990. Comparison of random search algorithm and mixed integer programming for solving area-based forest plans. *Canadian Journal of Forest Research* 20: 934-942.

Nelson, J.D. and Finn, S.T. 1991. The influence on cut-block size and adjacency rules on harvest levels and road network. *Canadian Journal of Forest Research* 21: 595-600.

Öhman, K. and Eriksson, L.O. 1998. The core area concept in forming contiguous areas for long term forest planning. *Canadian Journal of Forest Research* 28: 1032-1039.

Ohman, K. 2000. Creating contiguous areas of old forest in long term forest planning. *Canadian Journal of Forest Research* 30: 1817-1823.

Öhman, K. 2001. Forest planning with consideration to spatial relationships. (Thesis). *Acta Universitatis Agriculturae Sueciae Silvestria* 198. 32 pp.

Öhman, K. and Eriksson, L.O. 2002. Allowing for spatial consideration in long term forest planning by linking linear programming with simulated annealing. *Forest Ecology and Management* 161: 221-230.

Ohman, K. and Lamås, T. 2001. Clustering of harvest activities in multi objective long-term forest planning. In Forest planning with consideration to spatial relationships. (Thesis). *Acta Universitatis Agriculturae Sueciae Silvestria* 198: 1-15.

Richards, E.W. and Gunn, E. A. 2000. A model and tabu search method to optimize stand harvest and road construction schedules. *Forest Science* 46(2): 188-203.

Sessions, J. 1992. Solving for habitat connections as a Steiner network problem. *Forest Science* 38(1): 203-207.

Siitonen, M. and Nuutinen, T. 1996. Timber production analyses in Finland and the MELA system Large-scale forestry scenario models: experiences and requirements, Joensuu, Finland, 15-22 June, 1995. *EFI proceedings* 5: 89-98.

Snyder, S. and Revelle, C. 1996a. Temporal and spatial harvesting of irregular systems of parcels. *Canadian Journal of Forest Research* 26: 1079-1088.

Snyder, S. and ReVelle, C. 1996b. The grid packing problem: Selecting a harvesting pattern in an area with forbidden regions. *Forest Science* 42(1): 27-34.

Snyder, S. and ReVelle, C. 1997. Dynamic Selection of harvests with adjacency restrictions: The share model. *Forest Science* 43(2): 213-222.

Tarp, P. and Helles, F. 1997. Spatial optimization by simulated annealing and linear programming. *Scandinavian Journal of Forest Research* 12: 390-402.

Temple, S.A. 1985. Predicting impacts of habitat fragmentation on forest birds: a comparison of two models. In: Verner, J., Morrisson, M.L. and Ralph, C.J. (Eds.). Wildlife 2000: Modelling habitat relationships of terrestrial vertebrates. University of Wisconsin Press, Madison. pp. 301-304

Torres-Rojo, J M. and Brodie, J.D 1990. Adjacency constraints in harvest scheduling: an aggregation heuristic. *Canadian Journal of Forest Research* 20: 978-986.

Törnquist, K. 1996. Ecological landscape planning in swedish forestry. Integrating environmental values into forest planning, 27 June-2 July. Estonia, *EFI proceedings* 13: 189-196.

Weintraub, A. and Cholaky, A. 1991. A hierarchical approach to forest planning. *Forest Science* 37(2): 439-460.

Williams, J.C. 1998. Delineating protected wildlife corridors with multi-objective programming. *Environmental Modeling and Assessment* 3: 77-86.

Zipper, W.C. 1993. Deforestation patterns and their effects on forest patches. *Landscape Ecology* 8(3): 177-184.

Chapter 8

MEASURING NON-WOOD FOREST OUTPUTS IN NUMERICAL FOREST PLANNING

A review of Finnish research

TIMO PUKKALA

University of Joensuu, Faculty of Forestry, P. O. Box 111, 80101 Joensuu, Finland

Abstract. This chapter reviews some alternatives for numerically measuring the amount of forest outputs other than timber or economic profit. A common feature of the presented methods is that they can be used in numerical optimisation, either as a component of the objective or penalty function, or as a constraint. The chapter classifies the approaches for dealing with non-wood forest outputs into three categories, namely economic approach, numerical optimisation, and multi-attribute utility theory. The reviewed models devised for non-wood outputs are applicable to the numerical optimisation and utility theoretic approaches. The chapter gives several examples of both empirical and expert models, which have been developed in Finland to predict scenic beauty, amount of forest berries and mushrooms, and ecological quality of a forested landscape. The emphasis, in the description of ecological measures, is on variables, which help to mitigate the fragmentation problem of forest landscapes. In addition to models and approaches, the chapter also provides planning examples that utilise the discussed numerical models for non-wood outputs.

Keywords: biodiversity, expert model, forest amenities, landscape ecology, landscape planning, multiple-use, non-wood forest products, scenic beauty.

Timo Pukkala (ed.), Multi-objective forest planning, 173–207.

1. INTRODUCTION

The use of forests has traditionally been multi-objective in most cases. Forests have been used for hunting, as a source of firewood and construction materials and for collecting various non-wood forest products such as rattan, resin, berries and mushrooms. Forests also protect soil from erosion, regulate water resources and provide various amenities. Some forests have cultural and religious significance. However, especially in industrialised countries, this traditional mode of forest use has at least temporarily changed into almost pure timber production or profit making: one product has dominated forest use and management with a consequence that the forest's ability to provide other products and services has declined.

There are several reasons for this development. One is that a single goal has been considered so important that there is no need to consider other goals. It may also have been thought that non-wood forest benefits are produced in sufficient quantities even though they are not specifically considered in planning. One more possible reason is that a single authority, interest group or decision maker has dominated forestry decision-making, neglecting the opinions of other stakeholders.

But there are also theoretical and technical reasons. Many of the earlier methods developed for forest planning were able to deal with timber production only, and with computationally simple problems. Examples of these methods are the various forest regulation methods, or cutting budget formulas, which are still used in many countries. They help in finding a management for the forest that guarantees an even and sustainable flow of timber. Sustainability of timber production may be good for non-timber uses as well, at least better than an exploitative use, but it does not guarantee that the flow of other products and services does not deteriorate. Moreover, sustainable and even flow of timber may no longer be a relevant pursuit in all places.

During the past three or four decades forest scientists have demonstrated how to use mathematical programming, such as linear programming and goal programming, in forest planning (Dykstra 1984, Gadow and Puumalainen 2000). In cases where the nature of the planning problem has violated the assumptions of mathematical programming, various heuristics have been introduced that numerically seek good solutions to forest planning problems (Chapter 6). Numerical methods enable more efficient and more detailed timber production planning than simple regulation methods. They also enable simultaneous consideration of multiple objectives, for instance, through constraints, multi-objective utility functions, or penalty functions. However, now a new bottleneck emerges: mathematical methods and heuristics require that all goals must be expressed numerically. The lack of numerical measures for several goal variables has greatly reduced the possibilities for using numerical search and optimisation methods in multiple-use forest planning.

This chapter reviews several methods and models that numerically measure forest management objectives other than timber production or economic profitability. Some earlier-published models are repeated to illustrate a modelling approach or a model type. The article also includes a few imaginative planning examples based on real forests. The purpose of the examples is to illustrate and demonstrate the use of a model or a measurement approach in actual forest planning. Although this chapter briefly reports on some multi-objective case studies, the emphasis is on the quantification of a single goal at a time. Simultaneous measurement of several goals is dealt with in the other chapters of this book.

Two categories of forest management objectives other than timber production are reviewed in this chapter: multiple uses and ecological goals. They reflect the recent history of forestry and forest planning in Finland (Oksanen-Peltola 1999). For a long time the emphasis of scientific forest planning and research was on timber production. This era was followed by multiple-use forestry, which in Finland was mainly in the 1980s (Oksanen-Peltola 1999), the multiple uses being hunting, outdoor recreation, and collecting berries and mushrooms. Current thinking and forestry legislation also stress the importance of ecological objectives and nature conservation in managed forests. Ecological planning, or the combination of ecological and economic planning, is currently changing towards planning for sustainable development, in which social matters, participation and customary uses of forests are given a high priority.

2. MEASUREMENT OF FOREST OUTPUTS – A REVIEW OF APPROACHES

2.1 Categorising forest outputs

The benefits or values that a forest produces are many and they can be classified in several ways. One classification refers to products and services, or goods and services. Products are extracted or harvested from the forest whereas services do not need any harvesting. Timber is a forest product, and nature protection and scenic amenities are examples of services. Measurement of the amount of a product is straightforward and non-problematic. The forest's ability to produce services is related to forest features, which means that measuring the amount of a particular service is close to observing some relevant forest characteristics. Sometimes it is difficult to tell whether an output is a product or a service. For instance berry collection and hunting are mainly recreational activities in many countries, and their value might be much greater than the market value of the collected products, i.e., the berries or hunted game animals.

It is also possible to speak about forest uses or forest-related activities. Examples of forest uses are timber production, recreation and hunting. The total benefit of a particular use may come from several goods and services. Therefore, one way to evaluate a plan is to measure the value of each use, on the basis of the estimated amounts of those goods and services that are related to a particular use. The problem is that different uses are overlapping, for instance outdoor recreation cannot be distinctly separated from hunting and berry collection activities.

One more way is to talk just about decision criteria, which are specified by the decision-makers of a particular planning situation. The decision criteria may be forest uses, goods or services, depending on the situation. Each criterion is measured using characteristics –in this context called indicators or objective variables– which describe the quantity of a product or state of a particular forest feature. The criteria and indicators are defined separately for every planning and decision-making situation. When not dealing with a specific planning situation we may simply talk about outputs and inputs, or benefits and costs. The outputs and inputs are measured with variables that serve as indicators in planning and decision making.

2.2 Planning approaches

Economic approach

There are several approaches that include non-wood forest outputs in numerical forest planning. We can consider, for instance, economic or monetary measurement, numerical optimisation and utility theoretic planning. The economic approach tries to measure the monetary value of all forest outputs using various direct and indirect measurement methods. Measurement of outputs other than market products and market services is the most problematic part of the economic approach. How do we measure the value of recreation if recreation in the forest is free of charge? And how do we measure the value of a high biodiversity? Some estimation methods that try to overcome the measurement problem include the travel cost method, contingent valuation method and the method of hedonic prices (e.g. Tyrväinen 1999).

The total value of a service or forest feature may be expressed as the sum of use value and existence value (Naskali 1991):

Total value = Use value + Existence value

The value of biodiversity is obviously mainly existence value: high biodiversity or the existence of forest-dwelling species is valuable although biodiversity or the species in question is never "used".

The use value may be partitioned into two components:

Use value = Value of current use + Option value

The option value refers to the possibility of using the forest for a given purpose in the future although there may be no plans for such a use. A forest with more options for the future is obviously more valuable than a forest which is good for one purpose only, if the value of the current use is the same for both forests.

The current use value may in turn be divided into the following components:

Current use value = Consumption value + Non-consumptive use value

The non-consumptive use value refers to a use that does not consume the resource, like looking at a beautiful forest landscape.

Some of the forest goods, uses or services may have a market price that usually accounts for a part of the total value. Therefore, one more way to split the total value of an output is to divide it into market and non-market values.

Possibilities for hunting and collecting forest berries and mushrooms are examples of benefits having several significant components: the market value of the "product" collected during these activities may be significant, but also the non-market value of the current use, the option value of the potential use and the existence value of the plant or game species are important. The share of these components clearly depends on the individual user.

Figure 1 gives an idea of how the total value of some forest uses divides into various components (Naskali 1991). Threatened species and scenic beauty have no market value at all; their value consists of recreation, existence, and option values. The total benefit from berry collection and moose hunting consists partly of the market price of the capture, but their recreational value is greater. Option and existence values also contribute to the total value of hunting and berry collection.

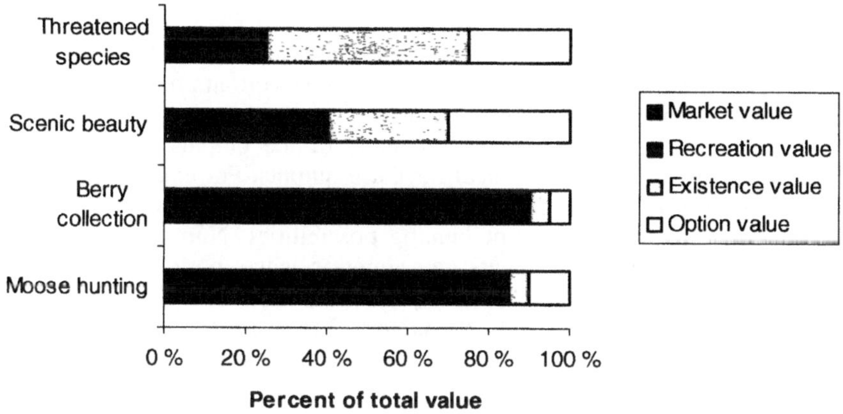

Figure 1. Components of the total value of some forest outputs.

If the value of an activity or type of forest use can be expressed as a function of the surface area of a certain type or forest, distance from a town, tree species composition, etc., it is possible to use this function in numerical optimisation, or apply marginal analysis to find a combination of forest uses that maximises the total monetary value of the combination of forest uses.

The economic approach to measuring non-wood forest benefits is very concrete because there is only one measurement unit familiar to all. However, despite these benefits, economic measurement might not be the best method, and not even a feasible one, for routine forest planning. The reason is that the measurement of the economic value in every planning situation is quite tedious. The estimates for the monetary value of a particular use, obtained with contingent valuation, hedonic pricing or other methods, often describe only a part of the total value. The results and relationships obtained in one survey may not be valid in another planning situation. Studies on the economic value often reflect common opinion. In forest planning, the estimates should indicate the value of the output to the parties that represent the decision makers of that particular planning situation. The economic approach also becomes quite complicated in situations where the same forest stands simultaneously produce several types of outputs, which is often the case.

Numerical optimisation

Another approach for dealing with non-wood outputs is numerical optimisation such as linear programming, goal programming or various heuristics. These methods do not require that the different outputs are measured with the same unit. Timber production can be measured by the volume of different timber assortments or the net present value of sold timber, and hunting possibilities by the population density of the game species or the surface area of suitable habitats. The value of berry collection may be measured by berry yield predictions (expressed in tons or kilograms) and scenic amenities by a scenic beauty score. No attempt is made to measure different outputs with the same unit. The variables are rather used to indicate to what extent a certain decision criterion is met. It is left to the decision maker to make the trade-offs and evaluate which decision alternative produces the best or most acceptable combination of outputs.

If there are no prediction models for some of the outputs, it is possible to consider characteristics that are critical for these outputs. For example, if a certain game species lives in old hardwood forests, it is possible to use the area of old hardwood stands as an indicator of hunting possibilities. Numerical optimisation does not require that the components (consumption value, option value, existence value, etc.) of an output's total value are specified or separated. What is relevant is that the decision maker understands the numerical measure and can use it to value the output. For example, the benefits of berry collection may be evaluated based on predicted berry yields, although only a small fraction of the yield is actually collected and the value of berry gathering is only partly related to the market value of the collected berries. It is assumed that the yield prediction simultaneously indicates the market value, consumption value, option value, and existence value.

Multi-attribute utility theory

The third approach that measures non-wood forest outputs is utility theory. This approach resembles the economic approach in that all outputs are converted into a single unit, which in this case is called utility (in fact, units disappear in this conversion). The original units of multiple outputs are first converted into sub-utilities, ranging from zero to one, and the sub-utilities from different outputs are then converted into a single total utility. The approach is also not distinctly different from numerical optimisation because the utility function can be used as an objective function in numerical optimisation, and the same indicator variables that are used in numerical optimisation can also be the variables that appear in the utility function. The utility function is developed separately for every planning situation. It is situation- and decision-maker-specific, and describes the importance of different management goals from the decision-maker's point of view. It also tells how different outputs can be traded off.

A combination of numerical optimisation and utility theory is a much-used approach to multi-objective forest planning. There are three basic ways to combine these approaches. The first way applies utility theory prior to optimisation, developing a utility function to be numerically maximised. The second way applies utility theory to prepare a model that relates the stand's or forest's priority in a certain respect to measurable forest characteristics. This priority model is not directly maximised but used in calculations in the same way as regression models, i.e., to calculate values of indicator variables (Kangas and Pukkala 1996, Kangas and Mononen 1997). The third approach uses numerical optimisation to generate alternative plans, producing different combinations of outputs and obtained by varying the problem formulation (e.g., by changing the constraints of a linear programming problem). The Analytic Hierarchy Process (Saaty 1980) or some other multi-attribute selection method based on utility theory is used after optimisation to compare the alternative plans (see Chapter 3). The calculated or subjectively assessed amounts of outputs are used as criteria and indicators in the decision analysis model.

2.3 Types of models for non-wood outputs

Empirical models

Numerical optimisation requires that the different outputs are expressed numerically ("quantified"). As planning deals with the future, the numerical measure is typically some kind of model prediction, or includes a model prediction. The most common type of model used in forest planning is the empirical statistical model, which is based on measured quantities of the output variable in question, and some independent (explanatory, preceding, predictor) variables. Empirical models are the most logical choice for most forest outputs. Berry yield measurements can be used to model berry yield as a function of known variables, and moose population counts to model the relationship between forest features and the density of moose population.

The relationships that determine the berry yield, density of moose population or some other non-wood output may be very complicated. However, a 'complete' or 'perfect' model is not needed. A model used in forest planning should contain information on the effect of those factors which are under the control of the decision maker and which are known in planning calculations. Weather conditions of the current year or flowering of the berry plant are useless predictors in a model, which is used to predict berry yields 10 years from the present.

Expert models

Unfortunately, empirical data are lacking for many non-wood forest outputs and collecting such data is often tedious and expensive. However despite lack of exact measurements, interested parties may have clear opinions or educated information about the relationships between specific forest features and the output in question. It is possible to convert this knowledge into a so-called expert model.

Several versions of expert models have been developed recently for forest planning purposes. Kangas and Mononen (1997) used the Delphi technique to develop a model that expresses forest biodiversity as a function of deadwood volume, area of old forests, and volume of broadleaved tree species. Eleven experts gave their opinion on the contribution of these factors to the overall biodiversity and on the priority of different levels of each factor. The results of the first query were summarised and sent back to the experts who were asked to express their opinions again, after seeing the overall opinions of all experts. The final model was developed from the results of the second query. The resulting model enables the use of a biodiversity score as an objective or constraining variable in numerical optimisation. The model looks as follows:

$$BD = 0.424\, p_1(\text{old forest}) + 0.250\, p_2(\text{broadleaf}) + 0.326\, p_3(\text{deadwood}) \quad (1)$$

where p_1, p_2 and p_3 are sub-priority functions expressing the priority of different amounts of the proportion of old forest (%), proportion of broadleaved species (%), and volume of deadwood (m^3ha^{-1}), respectively.

Another type of expert model uses paired comparisons to estimate the priority of different forests or stands with respect to a given output. A statistical model is then fitted to relate this priority to variables known in forest planning. As an example, Kangas et al. (1993a) asked 15 experts to make paired comparisons of 10 forested landscapes with respects to the needs of black grouse (*Tetrao tetrix*, *Lyrurus tetris* L.) using a verbal scale ranging from equal priority (1:1) to absolute priority of one landscape over the other (9:1 or 1:9). The verbal answers of one expert at a time were converted into numerical ratios (1:9, ..., 1:1, ..., 9:1) from which the priorities of forest landscapes were calculated, using the same eigenvalue technique as in the Analytic Hierarchy Process (Saaty 1977, 1980). In the next step, a model for the priority of a landscape was fitted using regression analysis. The predictors that gave the best fit described the amounts of birch, young sapling stands, and pine stands:

$$\ln(P) = -9.991 + 0.946 \ln(Birch) + 1.439 \ln(Height_{5\text{-}15}) + 0.023 \ln(Pine) \quad (2)$$

where P is the landscape's priority with respect to black grouse, *Birch* is the proportion of birch in the whole forest (% of standing volume), $Height_{5\text{-}15}$ is the proportion of stands in which the mean tree height is 5 – 15 m, and *Pine* is the proportion of stands (%) where the share of pine is at least 40 % of stand volume. The model explained 98 % of the variance of the mean priority (mean of 15 experts) of landscape.

The problem of this technique for deriving the priorities of observations (in the above case the 10 landscapes) is that the number of paired comparisons increases sharply as the number of observations increases. The total number of pairs of 10 landscapes is 10 × 9 / 2 = 45, and with 20 landscapes it would already be 20 × 19 / 2 = 190. One way to avoid excessive number of comparisons is to divide the landscapes or stands into groups and carry out the comparisons within each group. One member of each group is then taken into a scaling set, the members of which are again compared pairwise. The priorities of the scaling set are used to scale the priorities calculated for the initial groups (Kangas et al. 1993b). However, this method suffers from the drawback that the reliability of all priorities heavily depends on the few comparisons made within the scaling set.

Instead of paired comparisons, direct rating can be used to express the priorities of stands or landscapes (Pukkala et al. 1988). This method does not suffer from an excessive number of comparisons. However, with direct rating it is often questionable to assume that the priority is measured on the ratio scale or interval scale as required by the regression analysis.

To reduce the number of comparisons and to overcome some other limitations of the Analytic Hierarchy Process, Alho et al. (2001) proposed a regression method for developing a priority model directly from the paired comparisons. This method does not require a full set of paired comparisons. The minimum requirement is that each forest or stand belongs to one pair. If $r(i,j,k)$ is the evaluation of expert k for the value of stand or landscape i relative to j, the dependent variable of the regression analysis is $y(i,j,k) = \log(r(i,j,k))$. The value of stand or landscape i is then expressed as a function of p explanatory variables $x_{i1}, x_{i2}, \ldots, x_{ip}$, as follows (Alho et al. 2001):

$$v_i = \exp(\mu + \beta_1 x_{i1} + \beta_2 x_{i2} + \ldots + \beta_p x_{ip}) \tag{3}$$

Now the regression model can be formulated as follows:

$$\begin{aligned} y(i,j,k) &= (\mu + \beta_1 x_{i1} + \beta_2 x_{i2} + \ldots + \beta_p x_{ip}) - (\mu + \beta_1 x_{j1} + \beta_2 x_{j2} + \ldots + \beta_p x_{jp}) \\ &= \beta_1(x_{i1} - x_{j1}) + \beta_2(x_{i2} - x_{j2}) + \ldots + \beta_p(x_{ip} - x_{jp}) + \varepsilon(i,j,k) \end{aligned} \tag{4}$$

where $\varepsilon(i,j,k)$ is an error term with expectation zero. The intercept term μ cancels out. Recently, Ihalainen et al. (2002) used this approach to develop models for calculating a tree stand's priority for cowberry or bilberry collection. Silvennoinen et al. (2001) used the same technique for preparing models for landscape preferences at the forest stand level.

One question in the preparation of expert models is to decide who are the experts. Who is a person skilful enough to evaluate the suitability of a stand for a certain purpose? The answer clearly depends on the output being modelled. If we are modelling biodiversity or habitat quality of a threatened species it is reasonable to recruit experts who have proper training and sufficient professional experience. When modelling the habitat quality of game species, or the priority of stands for collecting non-wood forest products, the hunters or the collectors may be regarded as experts. At the other extreme, when modelling scenic beauty, which is a personal preference, every person may be an expert, especially when the model is used in a person's own forest.

Common, group and individual models

Empirical statistical models give objective predictions while expert models predict subjective preferences. Expert models are a way to automate subjective evaluations so that they can be made during numerical optimisation. An important aspect in expert models is to check that the model expresses the opinion of "right people". In this context we can refer to overall models, group models, and individual models. Overall models express the common opinion, or the consensus of the best experts, and they might be the best models in the planning of public forests. In private forestry, group or individual models may be used instead. Either a group model is estimated for a specific group, or group effects are included in the model (Silvennoinen et al. 2001). Individual models may be prepared for an individual (Kangas et al. 1993b), or a common model may be calibrated for a single decision maker.

As an example of group models, Silvennoinen et al. (2001) prepared the following prediction model for assessing landscape preferences at the forest stand level

$$\begin{aligned} v = {} & 0.04055\, H - 0.01284\, HG_1 - 0.02972\, HG_3 \\ & -0.009431\, N^{0.5} + 0.002565\, N^{0.5} G_4 \\ & + 0.01076\, S \\ & + 0.002725\, V_p P - 0.001476\, V_p P G_1 - 0.001863\, V_p P G_2 \\ & +0.0006877\, V_b B - 0.0005377\, V_b B G_2 + 0.001450\, V_b B G_3 \end{aligned} \tag{5}$$

where v is the priority of the stand with respect to scenic beauty; H is the basal-area-weighted mean height of trees (m), N is number of trees per hectare, S is transformed skewness of the height distribution, V_p volume of pine (m^3ha^{-1}), V_b is volume of birch (m^3ha^{-1}), and P and B are dummy variables. $P = 1$ if the basal-area-weighted mean height of pine is 10 m or more, otherwise $P = 0$. $B = 1$ if the basal-area-weighted mean height of birch is 10 m or more, otherwise $B = 0$. G_i is equal to one if the person belongs to group i, otherwise it is equal to zero (groups 1 and 2 refer to women, groups 3 and 4 to men, groups 1 and 3 to persons whose household owns forest, and groups 2 and 4 to persons whose household does not own forest). The regression coefficients indicate, for example, that the volume of large pines increases the priority clearly less with women (groups 1 and 2) than with men. The mean tree height increases priority much less with forest owners (groups 1 and 3) than with the non-owners.

Kangas et al. (1993b) proposed a method for modelling individual landowners' landscape preferences. The method is based on paired comparisons of stand photos, which are converted into priorities, which in turn are predicted from stand characteristics. The tests with 30 persons showed that most individual models correlate strongly with a general model (Pukkala et al. 1988), which predicts the mean priority of 121 people, but some individuals have landscape preferences which are distinctly different from the overall preferences.

3. MEASURES FOR FOREST OUTPUTS

3.1. Scenic beauty and other amenities

Scenic beauty is an important feature of the forest. It may be regarded as an output by itself, or as an indicator other benefits. Scenic beauty correlates strongly with the quality of the forest as a recreation environment (Pukkala et al. 1988). In Finland, scenic beauty of the forest is a very common secondary objective among individual forest landowners. For almost one third of owners the scenic beauty and other amenities are the primary goals of forest management (Ihalainen 1992) and it is reasonable to assume that their importance is increasing (Ihalainen 1990, 1992, Ovaskainen and Kuuluvainen 1994). Some public forests used for recreation may be managed primarily for scenic beauty.

From the measurement point of view, scenic beauty can be partitioned into different aspects or categories as follows:

- Beauty of a distant view
- Beauty of a within-forest view
    - Beauty of a single stand
    - Within-forest beauty of a forest area

Measurement or modelling the beauty of a distant scene is a most difficult if not impossible task, although some attempts to model it have been made (Shafer and Brush 1977). The most promising tool for evaluating the scenic beauty of future landscapes produced by alternative forest plans may be computer graphics, which are able to produce visualisations almost as realistic as photographs. The problem with the evaluation of visualisations is that the method is not amenable to numerical optimisation.

Within-forest scenic beauty of a single stand

Although distant vistas may produce the most famous and impressive forest scenes, only a rather small percentage of Finnish forests is significant in this respect. This is because forest terrains are flat, and many stands never belong to a distant view (if flight vistas are excluded). The relevant question in most cases is whether the forest is beautiful, representing a good environment for outdoor recreation when the visitor is in or near the forest. The approach used to measure this property is to first model the scenic beauty of an individual stand, as a function of growing stock characteristics, and then use the predicted stand-level scenic beauty scores to derive forest-level measures.

An example of a stand-level scenic beauty model is the following equation (Pukkala et al. 1988).

$$SB = 4.471 + 0.06450 d_g - 0.0001745 N + 0.006439 V_{Pine} D + 0.005733 V_{BirchAspen} D \qquad (6)$$

where d_g is the basal-area-weighted mean diameter of trees (cm), N is number of trees per hectare, V_{Pine} is volume of pine (m^3ha^{-1}) and $V_{BirchAspen}$ is volume of birch and aspen (m^3ha^{-1}). D is a dummy variable which is equal to one if the stand dominant height is 10 m or more, and zero otherwise. The function predicts that sparse stands of large trees, preferably pines and birches, are experienced as beautiful.

The above model for a stand's scenic beauty is based on slides of 100 different stands. A total of 121 people rated the stands using a scale from 0 (extremely poor) to 10 (extremely good). From these evaluations, the mean score was calculated for every stand, and the mean score was then predicted from measurable growing stock characteristics. The model explained 69 % of the variation in a stand's mean scenic beauty score. It turned out that respondents belonging to different categories (forest owners, foresters, forestry students, other students, school children, 'ordinary citizens') had quite similar landscape preferences. It also turned out that the stand's recreation score (priority as a recreation environment), which was also obtained from the respondents, correlated very closely with the scenic beauty score.

The above models were prepared in 1988. About a decade later, Silvennoinen et al. (2001) prepared a corresponding scenic beauty model, but the stand photos were evaluated using paired comparisons of stands instead of direct ratings. A model for stand's scenic beauty was fitted, using the regression version of the Analytic Hierarchy Process (Alho et al. 2001). The new model, estimated with an improved technique, gives results very similar to the old one, indicating that Finns' landscape preferences have not changed in the recent past.

Within-forest scenic beauty of a forest area

If the decision maker of a particular forest planning situation is interested in scenic beauty or recreation, the objective is usually related to the whole forest and not to individual stands. There are several possible ways to compile forest-level measures from the stand level scores or priorities (Pukkala et al. 1995). Examples of numerical forest-level within-forest scenic beauty measures are:

- Area-weighted mean score of stands
- Location- and area-weighted mean score of stands
    - Location weights depend on visibility
    - Location weights depend on accessibility or proximity
    - Location weights are subjectively determined
        - by specialists
        - by ordinary people

GIS packages enable various analyses that produce visibility and accessibility estimates for forest stands. The visibility figure tells how often the stand is seen from points that people often visit, and accessibility measures the distance of a point or stand from various entry points such as roads, trails, hotels and cottages (Pukkala et al. 1995). The results of GIS analyses and stand-level priority models can be used to calculate visibility- or accessibility-weighted means of scenic beauty scores, which are usable as objective or constraining variables in optimisation. The weights are static if visibility and accessibility are calculated without taking the tree cover into account, which means that there is no need for any spatial calculations during the optimisation; a location-weighted mean of a stand variable can be used as an objective or constraining variable in, for instance, linear programming and goal programming.

The viewpoints or entry points, from which the visibility or accessibility of a location is measured, are usually subjectively placed. The location weights of stands are therefore also partly subjective. An interesting approach used to derive subjective weights for locations is to ask people, for instance via Internet, to choose places in the forest that they think are important for recreation or nature conservation (Hytönen et al. 2002). The responses may be compiled into a "wish map", which more or less directly gives the location weights to be used in planning.

If it is assumed that the variability of the forest improves its quality as a recreation requirement, it is possible to maximise the variability of some stand characteristic simultaneously with the scenic beauty or recreation score. For example, Pukkala et al. (1995) maximised a utility function, the value of which depended on the accessibility- and area-weighted mean recreation score of stands, and the length of a compartment boundary that separates two stands differing at least 4 meters in mean tree height. The latter objective was based on the assumption that a person moving through the forest finds it more interesting (less monotonous) when the stand type changes every now and then (Gustke and Hodgson 1980, Nousiainen and Pukkala 1992).

Zoning

As well as for deriving location weights, the results of visibility, accessibility and proximity analyses, or wish maps, may be used for zoning, i.e., for pointing out compartments or parts of compartments which are visible, often visited or otherwise in sensitive places (Store 1996). The resulting zones may be utilised in two ways in planning calculations: sensitive zones are given management objectives and constraints different from the rest of the forest, or the treatment alternatives simulated for sensitive areas are different from those for the other areas (Nalli et al. 1996, Næsset 1997a, 1997b).

Although zoning of the forest for different purposes with different management objectives and guidelines in each zone sounds like a clear and straightforward procedure, the resulting plan may be less efficient than a plan created without strict zoning. This is because the zones and their management alternatives are subjectively derived (not optimised). In addition, it may not be optimal to divide the forest into distinct zones dedicated to different uses but to use all parts of the forest for all uses with weightings determined by optimisation.

Zoning based on optimisation may be termed soft zoning because the zone boundaries are vague, change in time, and cannot be drawn on a map. Distinct zones dedicated to different forest uses represent hard zoning. Hard zones are an example of exogenous planning while soft zoning represents endogenous planning (Siitonen 1999, Kurttila 2001, Öhman 2001). In the endogenous approach the decisions related to zoning are proposed in optimisation while in the exogenous approach they are exogenously given, i.e., decided beforehand. The advantage of the endogenous approach is that a great number of alternatives –including those for zoning– are evaluated, and the trade-offs between different management goals are analysed. This analysis is impossible in the exogenous approach (Hof and Bevers 1998).

Forest-level example

A small example illustrates the effect of using forest-level scenic beauty objectives in numerical optimisation. Three 30-year plans were prepared for a forest area of 981 hectares. All plans had a constraint that exactly 150 000 m^3 must be harvested during the 30-year period. Plan 1 maximised the remaining growing stock volume with the 150 000 m^3 harvest constraint. Plans 2 and 3 had scenic beauty at the end of the 30-year period as the second objective, equally important as the growing stock volume. Plan 2 used the area-weighted mean scenic beauty score of stands while Plan 3 used a stand's visibility from a road passing through the area as an additional weight in the calculation of mean scenic beauty score (Fig. 2). The location weights were calculated by placing viewpoints on the road, then calculating the number of viewpoints which could be seen from different raster cells of the digital elevation model if terrain only were taken into account. Only viewpoints within 300 meters were included because weights were wanted for the within-forest (close scene) scenic beauty score. In the second step, the average visibility of each stand was calculated from the visibility counts of all cells that belonged to the stand, the result being the stand's location weight.

The first solution (Plan 1) uses the most clear-felling, much of which is next to the road along which the viewpoints are located (Fig. 2). Plan 2 cuts exactly the same volume as Plan 1, but thinnings are used more frequently and clear-fellings less so than in Plan 1. However, many clear-fellings are still near the road (Fig. 2). In Plan 3, which used a location-weighted mean scenic beauty score as an objective, none of the clear-fellings joins the road. Stands next to the road are treated with thinnings, and clear-fellings are located further away from the road.

Table 1. Values of some variables calculated for three 30-year forest plans for an area of 981 ha. All plans cut 150 000 m^3 during the 30-year planning period. Plan 1 maximises the remaining growing stock volume after 30 years while plans 2 and 3 have the mean scenic beauty score of stands (SB) as an additional objective variable. Plan 2 uses the compartment-area-weighted mean SB and Plan 3 calculates the SB using both the compartment area and the stand's visibility as weight variables.

Variable	*Plan 1*	*Plan 2*	*Plan 3*
Harvested volume (1000 m^3)	150	150	150
Remaining volume (1000 m^3)	155	155	155
Area-weighted SB	5.6	5.9	5.7
Location- and area-weighted SB	5.5	5.8	7.0
Clear felling area (ha)	302	257	273
Thinning area (ha)	165	304	268

Table 1 shows that the three plans are equally good in terms of harvested and remaining volume, but plans 2 and 3 produce more beautiful stands. The area-weighted mean scenic beauty score is best in Plan 2 but the location-weighted scenic beauty score is clearly the best in Plan 3. In practice, the stand scenic beauty score varies from 3 to 8, which means that a change from 5.6 to 5.9 is already a significant improvement, and the improvement of the location-weighted score from Plan 1 to Plan 3 (from 5.5 to 7.0) is especially great.

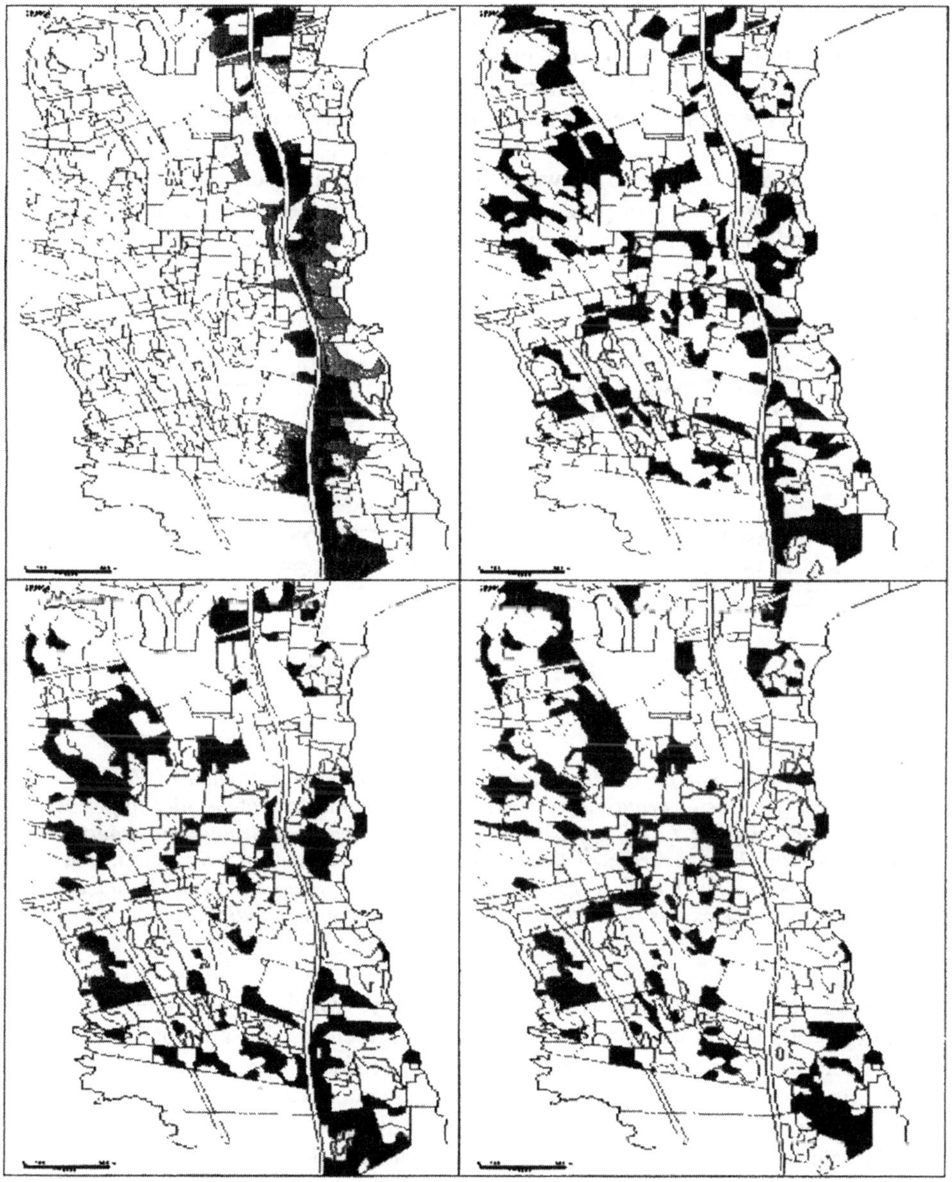

Figure 2. Top left: Stands' visibility from the road (dark colour implies high visibility). The other maps show the cutting areas. Black is clear-felling, dark grey removal of over-story trees and light grey a thinning treatment. Top right: Plan 1 (no scenic beauty objective); bottom left: Plan 2 (mean scenic beauty score of stands used as an objective); bottom right: Plan 3 (location-weighted mean scenic beauty score of stands used as an objective).

The above example suggests that improvements in the scenic beauty do not bring any losses in timber production goals. However, this is only partly true because, in the long run, longer rotations with lower average profitability will be used in areas where scenic beauty is important. This will decrease the overall profitability of timber production. Fortunately, numerical measurement of scenic beauty allows for numerical analyses of the trade-offs between scenic beauty and economic goals (Fig. 3). The analysis shown in Figure 3 indicates an increasing rate of transformation and a very convex trade-off curve between scenic beauty and economic profit. This suggests that the possibility of producing both benefits in the same area is good, and it is easy to find compromises between economic and scenic beauty targets.

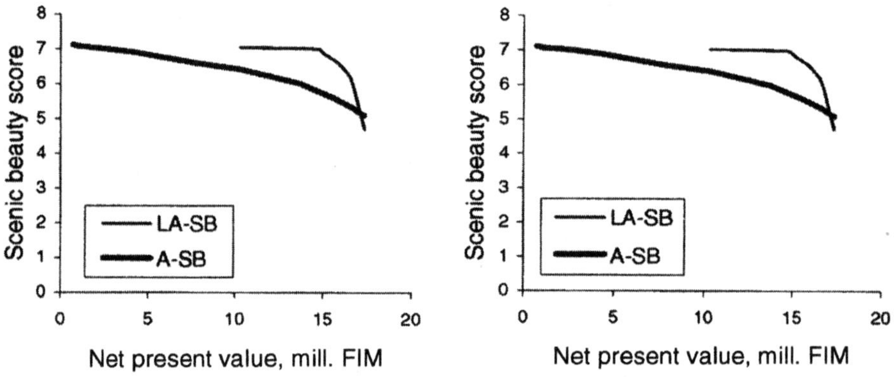

Figure 3. The production possibility boundary between the net present value of 30-year income and the mean scenic beauty score of the stands (SB) at the end of the 30-year period for a 981-hectare forest. LA-SB is the location- and area-weighted mean scenic beauty score, and A-SB is the area-weighted mean score. The curves on the left have a constraint that the soil expectation value of the forest at the end of the 30-year period must be at least the same as it is currently (reflecting economic sustainability).

The above examples concerned scenic beauty, which is just one class of amenities produced by the forest. Other amenities include, and are not limited to, outdoor recreation, hunting, and collecting forest berries and mushrooms. Hunting, or gathering berries and mushrooms, can be recreation or profit-oriented activities or a combination of both. When recreation is just walking or jogging in the forest, the forest's value for recreation correlates closely with the within-forest scenic beauty (Pukkala et al. 1988) and it can be measured in the same way as scenic beauty. Hunting benefits may be evaluated with the same techniques as the ecological quality of the forest (see Section 3.3).

3.2 Non-wood forest products

In many countries non-wood forest products (NWFP) form a very important category of forest outputs with an equally high or even higher economic value than timber. Examples of important NWFPs are resin, rattan, bamboo, berries, mushrooms and various medicinal plants. In countries where people no longer live in the forest, the collection of NWFPs is mainly a recreational activity. However, in the same countries, both the economic and recreational value of NWFPs may be very important in some planning situations and for some forest owners (Ihalainen 1992). Therefore, good planning systems should be able to predict the ability of forests to produce NWFP-related benefits in most if not all countries.

If the collection of non-timber forest products requires permits, the value of the activity may be measured through permits. It is possible to model the willingness to pay as a function of some forest properties. This function may then be used in numerical optimisation to measure the hunting or collection value of different solutions. However, whenever hunting or berry collection are customary uses, this approach will certainly give biased results, besides being a tedious method to apply in a single planning situation.

Most NWFPs are related to the growth and yield of plants in the same way as timber production depends on the growth of trees. Measuring NWFP may therefore be approached through modelling the growth and development of the plant species which produce the NWFP in question. The problem of this approach is that collecting the input data for these models in forest inventory may be too tedious, and the modelling data are often lacking. Therefore, a more realistic approach is to model the relationship between the yield of a NWFP and the site variables and growing stock characteristics that are measured in the inventory and known in planning calculations. What is relevant in this pursuit is that the models express the effect of so-called controllable variables on the yield of NWFP. Examples of controllable variables are the density or species composition of the stand, and treatment of forest soil (soil preparation, fertilisation, ditching). Sometimes, as in the case of berries and mushrooms, the yield depends much more on the weather conditions of a particular year than on characteristics that can be controlled. This means that models using site and stand characteristics as predictors explain only a small fraction of the yield variation. However, this small part is all that planning needs to find the correct treatment for the stand.

Typical to many NWFPs is that their occurrence is clustered both temporally and spatially. In empirical modelling this means that large data sets should be collected or large sample plots used, and data should be collected over many years. This, in turn, makes empirical modelling based on measured yields very slow and costly. Therefore, other solutions, based on expert knowledge, are attractive in modelling NWFPs. Ihalainen et al. (2002) prepared models for cowberry and bilberry (the two most important forest berries of Finland) based on the evaluations of 100 stands by 27 experts including active collectors, biologists and foresters. The resulting models related the priority of the stand, as a berry collection place, to the site and stand characteristics as follows:

$$\ln(v_B) = 0.0062T_g - 0.0136G + 0.0363H_{dom} + 0.0014V_p - 0.0013V_d + 0.2393S \quad (7)$$

$$\ln(v_C) = 0.0053T_g - 0.0024V_p - 0.0033V_d + 1.6652S - 0.1673SD_g + 0.005SD_g^2 \quad (8)$$

where v_B is the predicted priority of the stand in terms of bilberry yield, v_C the predicted priority in terms of cowberry yield, T_g the mean age of trees (years), G the stand basal area (m^{2}ha^{-1}), H_{dom} the dominant height (m), V_p the standing volume of pine (m^{3}ha^{-1}), V_d the standing volume of deciduous trees, D_g the mean diameter of trees (cm), and S is a dummy variable which is equal to one if the forest site type is *Vaccinium* type (rather poor) or poorer, and zero otherwise.

Ihalainen and Pukkala (2001) also proposed a modelling approach that is an intermediate type between empirical and expert modelling. In this approach a surveyor rates the stand in terms of berry yield while measuring the stand characteristics in forest inventory. Berry yields are not measured, but the surveyor can use the density of the berry vegetation and the amount of berries to support the rating. The experiences from different modelling approaches are that the predictions of different models correlate positively and rather strongly, and in this way corroborate each other. For example, the bilberry model based on field ratings had a correlation of 0.797 with the expert model shown above, and the correlation between the two cowberry models was 0.689. However, it is impossible to tell which modelling approach is the best until large sets of empirical data are available.

3.3 Ecological objectives

Approaches

Ecological objectives in forest planning aim at retaining or improving the biodiversity of a forest. Forest biodiversity is retained if the populations of all forest-dwelling species remain viable. From the viability perspective, the species may be classified into three categories (Mykrä and Kurki 1998): (1) species which are not adversely affected by forest management, (2) species that may be threatened by forest treatments, and (3) species that can survive only in conservation areas. The key group in the management of commercial forests is group number two.

Because the number of species that may be adversely affected by forestry is certainly very high, it is not possible to measure separately the population viability of all species. The lack of ecological knowledge on species' requirements, in addition to an excessive number of species and thus an overly-complicated planning problem, prevents the full coverage of flora and fauna. Therefore, several simplified methods have been proposed for measuring the ecological quality of a forest area. These methods more or less clearly represent two distinct approaches: the ecosystem approach and the species approach (Fig. 4).

The ecosystem approach suggests that if a managed forest ecosystem maintains some key features of, or resembles, a naturally evolved ecosystem in some relevant respects, the species' viability is not threatened. The species approach specifies some keystone or umbrella species, the viability of which is evaluated individually. The advantage of the ecosystem approach is that it covers all species, but there is no certainty that the variables used to measure the ecosystem's ecological quality actually guarantee the population viability of a single species. The species approach, being more specific, may be better for maintaining the populations of the selected keynote species, but the viability of the keynote species does not necessarily guarantee that all the other species are also saved. In addition, the risk arising from using uncertain ecological knowledge may be higher in the species approach.

In forest planning, the future states of the forest are being evaluated. Therefore, a measure that describes the current status of the forest is insufficient for forest planning. For example, biodiversity assessments or plant species inventories are not enough for forest planning unless there are models for predicting the temporal change of the inventoried biodiversity characteristics and how the change is affected by forestry operations. However, assessments of the current forest may greatly help forest planning in specifying key locations and key species to be considered in planning calculations.

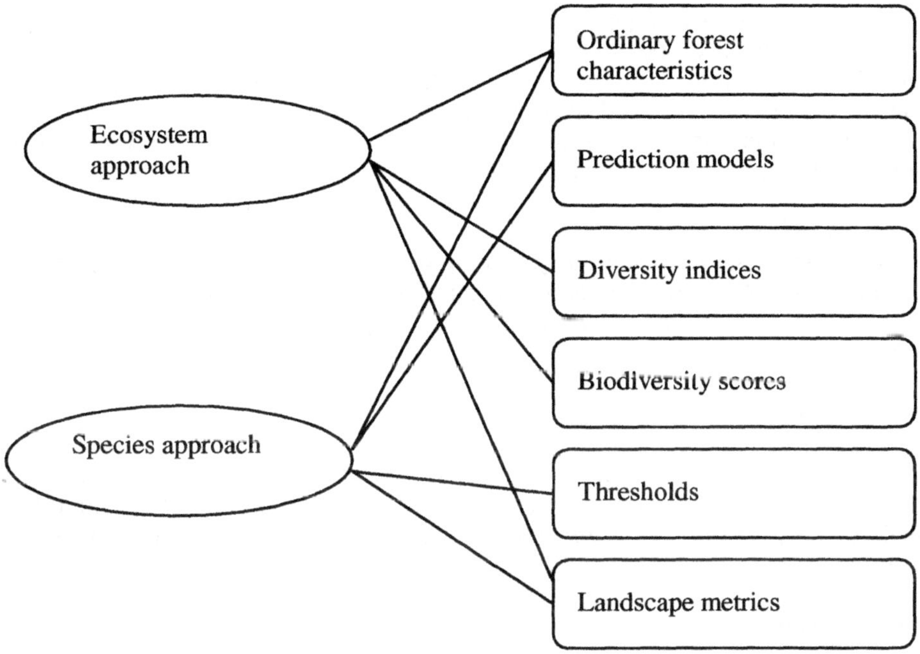

Figure 4. Alternative ways to measure the ecological quality of forestland in numerical forest planning.

Ordinary forest characteristics

Figure 4 indicates some ways to numerically measure the ecological quality of forestland. Ordinary forest characteristics refer to measurable variables which indicate high biodiversity and which are usually known in forest planning calculations. Examples of these are the area of forest older than a specified age, total volume of a certain tree species or species group, and volume of trees larger than a user-specified diameter. Some planning systems can also predict the accumulated volume of deadwood. When using these kinds of forest characteristics as objective or constraining variables, the user may think of the forest biodiversity as a whole, or from the standpoint of the viability of a single species.

Prediction models

An obvious tool for ecological planning is a model that predicts, as a function of measurable forest characteristics, the population size of a species, habitat suitability, or forest's priority with respect to this species. These models may be based on empirical data concerning population densities and forest features. As the territories of several species may be large, empirical data on both the forest and the species in question should also cover large areas. To model relationships, the data should cover several areas with different population densities and forest features. The fact that population sizes also depend on factors other than forest variables, such as predation, hunting, diseases and weather conditions, makes it difficult to reliably model the relationships between controllable forest characteristics and population viability. To overcome this difficulty, Kangas et al. (1993a) used expert modelling to derive relationships between measurable forest characteristics and the quality of forest as the environment of black grouse.

Diversity indices

Diversity indices, which are many in number, represent the ecosystem approach. Examples of indices are those of Shannon and Weaver, and Berger and Parker (e.g., Kouki 1993). The species diversity index of Berger and Parker (D') is computed from

$$D' = N_{max}/N \tag{9}$$

in which N_{max} represents the number of individuals of the most frequent species, and N the total number of plants or animals in the sample. Low value of D' indicates high species diversity. The index of Shannon and Weaver (H') is defined as follows:

$$H' = \sum_{i=1}^{k} \ln(p_i)p_i \tag{10}$$

where p_i is proportion of species i in the sample (N_i/N), and k is the total number of species in the sample. High value of H' indicates high species diversity.

When using indices one should remember that forest planning compares different future states of the same forest, and not the current state of different forests. This limits the use of indices based on species frequencies because the future frequencies cannot usually be predicted. However, what the planning system can predict are the future frequencies of tree species. Therefore, indices computed from tree species frequencies may have some use in the comparison of alternative forest management plans, especially if the flora contains a reasonably high number of species. The rationale in this kind of use is that high tree species diversity also indicates high overall diversity. A drawback of the approach is that planning systems can usually predict the future amounts of only economically important species with a reasonable accuracy. Also, there is not much empirical or theoretical evidence that the overall diversity indices reliably measure the viability of single species. As an alternative to tree species diversity, it is also possible to calculate the diversity index from the areas (≈"frequencies") of different stand types (or age classes, species compositions, etc.).

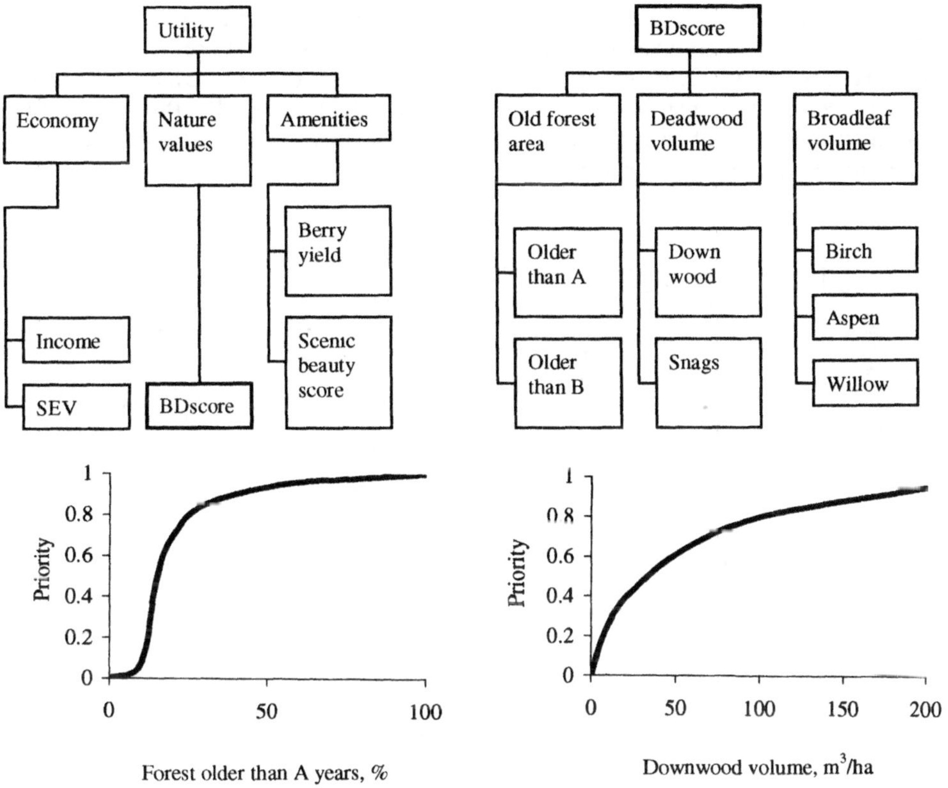

Figure 5. The principle of using a biodiversity score (top right) as a part of a decision model (top left). Bottom: examples of functions for calculating sub-priorities for the components of the biodiversity score.

Biodiversity scores

Biodiversity scores represent the ecosystem approach and calculate an overall biodiversity score based on selected criteria and indicators (Kangas and Pukkala 1996, Pukkala et al. 1997). The approach resembles the use of measurable forest characteristics to indicate biodiversity. It also resembles prediction models because the criteria are given weights that correspond to the coefficients of a regression model. The difference is that the criteria and indicators, as well as their weights, are specified beforehand by experts or with experts' help, and they are considered to be less situation-specific than the other decision criteria. The biodiversity score may be seen as a sub-hierarchy in the decision hierarchy, or as a sub-model of the utility function, developed for a particular planning case (Fig. 5).

Kangas and Pukkala (1996) proposed that such forest characteristics be used as the criteria and indicators of biodiversity that limit or reduce biodiversity. In Finland, lack of old forests, deadwood and some broadleaf tree species are often believed to reduce biodiversity. These forest features may therefore be taken as the biodiversity criteria. Each criterion may further be measured by several indicators, which are variables that the planning system can calculate. Finally, the priorities of different quantities of an indicator variable are described by a sub-priority function, determined separately for each indicator (Fig. 5).

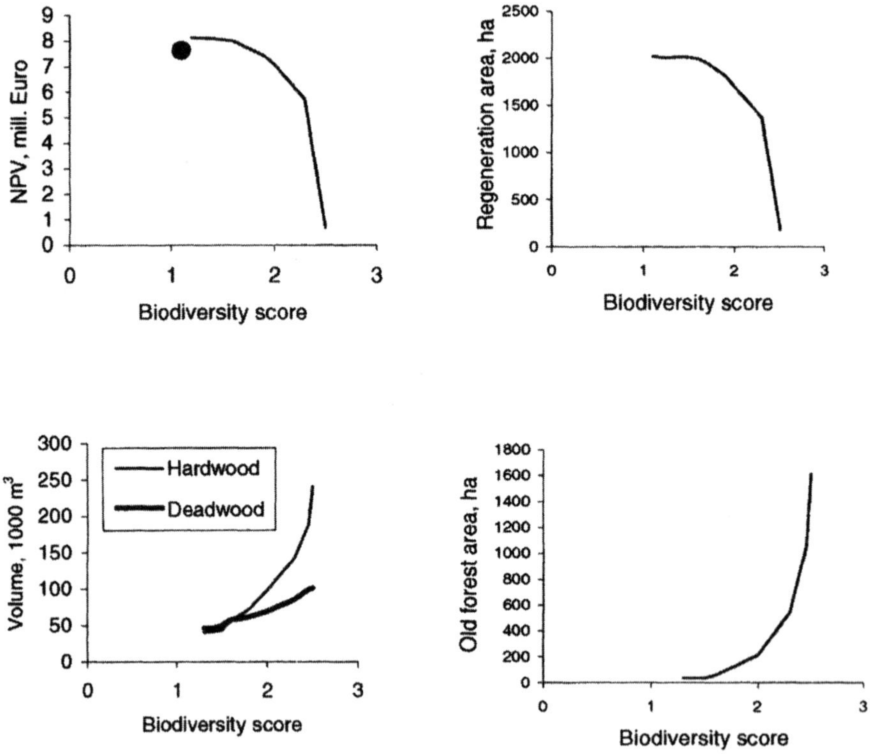

Figure 6. Top left: Production possibility boundary between discounted 30-year net income and biodiversity score of the final forest (after 30 years) in a 2400-hectare forest. The dot represents the solution that minimises the biodiversity score. The other diagrams show how the regeneration area and the components of the biodiversity score change when one moves to the right along the production possibility boundary.

A biodiversity score developed in this way may be used as an objective variable in numerical optimisation. An example calculation is presented in Figure 6 in which the relationship between discounted 30-year harvest income and biodiversity score of the final forest has been analysed in a forest area of 2400 hectares. The biodiversity score depends on the area of old forest, volume of dead wood, and volume of selected broadleaf species at the end of the 30-year planning period. The analysis indicates that the production possibility boundary between harvest incomes and biodiversity score is very convex representing an increasing rate of transformation between these management goals. At first, an increasing biodiversity score does not notably decrease the net present value of incomes. However, additional improvements in the biodiversity score become more and more expensive in terms of the other management goal, until almost all cuttings are excluded from the solution.

Thresholds

The threshold approach refers to the specification of limits for a resource or resources which are considered critical for certain keynote species. The thresholds may be based on expert knowledge or empirical data, or they may be determined using a model for population dynamics. A population model may be used to calculate the probability that the population will survive in different environments (Kangas and Kangas 1998). In these calculations, one of the parameters of the population model should depend on a measurable and controllable forest variable, i.e., a variable for which the threshold is being calculated. The results obtained with a population model may be used to specify a single threshold value (arrow in Fig. 7) or to develop a sub-priority function for the critical resource (solid line in Fig. 7). The threshold value can then be used in numerical optimisation, for instance as a constraint of a linear programming model. The sub-priority function may be used as a multiplicative component in the objective function, for instance as follows (Kangas et al. 2000):

$$P = \prod_{i=1}^{n} p_i(q_i) \left[\sum_{j=1}^{m} w_j u_j(q_j) \right] \tag{11}$$

where P is the priority of the solution, p_i is a priority function for species i and q_i is the quantity of the resource critical for species i. If the priority function for any species included in the objective function is zero, then the value of the whole plan also becomes zero, indicating that the plan is unacceptable. The latter part of the objective function ($\sum w_i u_i(q_i)$) is an additive utility function expressing the importance and contribution of the goal variables between which trade-offs are possible.

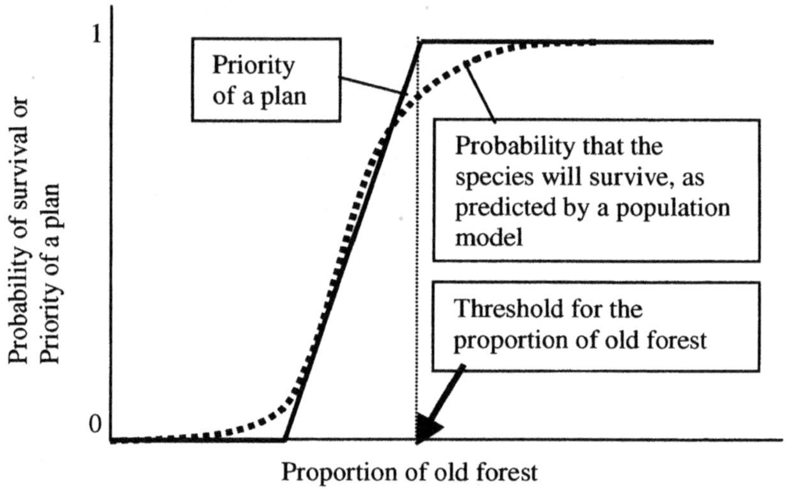

Figure 7. The idea of using a population model for specifying a threshold or a sub-priority model. A threshold is the minimum proportion of old forest that maintains a viable population with a high probability.

Landscape metrics

One more approach to measuring the ecological quality of a forest involves variables, here called landscape metrics, which measure the structure and composition of a forest landscape (McGarical and Marks 1995). Before calculating the metrics, the stands are classified into different habitat types or, if only one species is concerned, as habitats and non-habitats. It is also possible to develop a continuous stand-level habitat suitability index (*HSI*) which indicates the suitability of a single stand as habitat for a certain species. For example, Kurttila et al. (2002) developed the following equation for calculating a stand-level *HSI* for flying squirrel:

$$HSI = \prod_{n=1}^{N} (P_n)^{1/N} \tag{12}$$

where P_n is the sub-priority function for stand characteristic n related to habitat quality and N is the number of variables included in the *HSI*. Five stand characteristics were included in the *HSI*, namely growing stock volume (m^{3}ha^{-1}), proportion of Norway spruce in the total volume, proportion of deciduous trees, mean diameter of deciduous trees and volume of dead standing trees (m^{3}ha^{-1}). The maximum value of each sub-priority function was one (indicating that the amount of the variable is sufficient), which means that the maximum value of *HSI* is also one.

Once the stands have been classified into different habitat types, or habitats and non-habitats, many different indices can be calculated, some of which are non-spatial while others are spatial (McGarical and Marks 1995). If the habitat quality is measured with a continuous variable it is also possible to calculate, among others, the mean *HSI* of the forest, which is a non-spatial measure, or the spatial autocorrelation of the *HSI*. A few recently used examples from Finland (Kurttila et al. 2002) and Sweden (Öhman 2001) include:

- Non-spatial landscape metrics
    - Habitat area (or proportion of habitat)
    - Mean *HSI*
    - Location-weighted mean *HSI*
- Spatial landscape metrics
    - Habitat–habitat boundary
    - Habitat–non-habitat boundary
    - Core area (or proportion of core area)
    - Spatial autocorrelation of *HSI*

Practical considerations

In Finland and Sweden (Öhman 2001), the fragmentation of the habitats of key species is often thought to decrease species' abilities to sustain viable populations. Therefore, variables used to measure a forest's ecological quality should measure the aggregation and connectivity of habitat patches as well as their total area. This means that a justified ecological objective is often spatial in nature, and it depends on the locations of different types of forest. There are many ways to derive spatial objective variables, but when used in numerical optimisation, the calculation of the objective variable should not be too complicated. When iterative heuristics are used, it is preferable that the calculation of the variable does not require special software such as a GIS program during the optimisation run, because this would restrict the possibilities of optimisation. GIS analyses may be used before the optimisation to produce adjacency information or any other data required for the landscape metrics. Tens or hundreds of thousands of solutions, or candidate plans, are often evaluated during an optimisation run if the area is not very small. Optimisation basically consists of making changes in the solution, evaluating the objective function value and the feasibility of the solution after every change, and deciding about the next move. The time needed for the optimisation depends on two things. First, it depends on how many changes or moves the optimisation algorithm makes before finding a solution that is optimal or reasonably close to it. A good algorithm requires few objective function evaluations but is able to find an optimal or nearly optimal solution. The second factor is the time needed to calculate the effect of a change on the values of objective variables.

A change of the solution, or a move, usually means altering the treatment proposal of a single stand or compartment. In non-spatial optimisation a move's effect on the objective function depends only on the compartment in which treatment was changed. In spatial optimisation, a re-evaluation of the objective function requires an inspection of the neighbouring stands as well, and sometimes also neighbours of neighbours, etc. The greater the number of stands that must be inspected to re-evaluate the objective function, the slower the optimisation process will be.

A location-weight example

In a recent study, Kurttila et al. (2002) tested three different types of spatial objective: (1) location-weighted mean *HSI*, (2) spatial autocorrelation of *HSI*, and (3) the length of habitat–habitat (H–H) or habitat–non-habitat (H–NH) boundary. The purpose was to test alternative objective variable types, which have distinctly different features, are easily calculated during optimisation, and can be used for several species and also for non-ecological purposes. All three objective types can be used to cluster habitats or scanty resources; types 2 and 3 may also be use to disperse them.

The location-weighted mean *HSI* requires that the weights are calculated for different compartments so that they describe the importance of having a high *HSI* in a compartment (Store and Kangas 2001). The location weights are high in places where resource clusters are wanted: near existing old forest patches or current dwelling areas of the species, for instance. Figure 8 illustrates the effect of location-weighted mean *HSI* of flying squirrel on the optimisation result in a small sub-area of a 920-ha forest in an inventive planning case. The location weights are based on the proximity of observed nests of flying squirrel. Such a proximity analysis can be performed using standard GIS routines. Three plans were prepared using heuristic optimisation. All plans had a strict requirement that the harvested timber volume during a 30-year planning period must be at least 100 000 m^3. With this requirement, Plan 1 maximised the growing stock volume after 30 years. Plan 2 had the mean (area-weighted) *HSI* of the compartments at the end of the 30-year period as a second goal, of equal importance to the growing stock volume. Plan 3 used location- and area-weighted mean *HSI* as the second objective.

The three plans differ greatly in the sub-area shown in Figure 8. Compared to Plan 1, Plan 2 results in better *HSI* values (Table 2). In Plan 3, the important resource occurs in a less fragmented way than in plans 1 and 2. The area-weighted mean *HSI* is the same in plans 2 and 3, but the location-weighted mean *HSI* is clearly better in Plan 3 (Table 2). Having the habitat suitability index of flying squirrel as a management objective reduces the remaining volume and the clear-felling area, but the reductions are not large (Table 2).

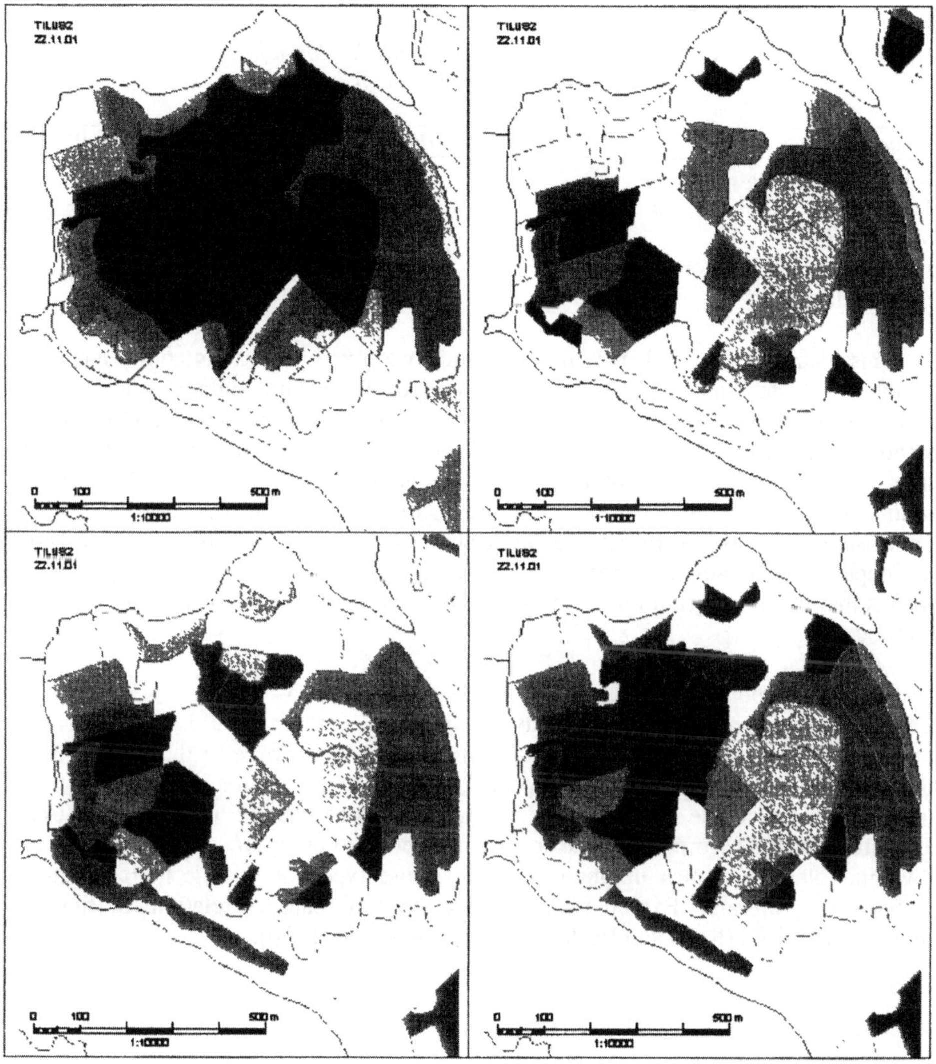

Figure 8. Top left: location weights (dark colour implies high location weight and white indicates a zero weight). The other maps show the value of flying squirrel habitat suitability index (HSI). Top right: no habitat goals; bottom left: the mean HSI was maximised together with other goals; bottom right: location-weighted mean HSI was used as an objective variable. In all HSI maps, darker colour implies higher habitat suitability index. White colour indicates an HSI equal to zero.

Table 2. Values of some variables in three 30-year plans developed for a 920-hectare forest. HSI is the habitat suitability index of flying squirrel.

Variable	*Plan 1*	*Plan 2*	*Plan 3*
Harvested volume, 1000 m^3	100	100	100
Total volume after 30 years, 1000 m^3	128	123	121
Mean *HSI*	0.200	0.208	0.208
Location-weighted mean *HSI*	0.165	0.215	0.322
Clear felling area, ha	285	271	264

This example shows clearly that a location-weighted habitat suitability measure is able to cluster habitat patches. In the absence of a habitat suitability index it is possible to use an ordinary forest characteristic, that measures the suitability of the stand for a particular species, such as stand age, volume of certain tree species, mean tree size, or deadwood volume. Static location weights, computed prior to optimisation, have the advantage that no spatial calculations are needed during optimisation because the effect of a single stand on the objective function value does not depend on the neighbouring stands. This means that optimisation is fast. It also means that linear programming and its modifications can be used as the optimisation method.

An example of spatial autocorrelation and habitat boundary

Location weights assume that the clusters of habitat patches should be in the same places continuously, i.e., they are not allowed to move over time. Although this may be a justified requirement for some species, it may also result in a less efficient plan than one in which the resource clusters are allowed to change location over time. Therefore, if it is assumed that the species can move to other areas when the current dwelling place becomes uninhabitable, objective types not using static location weights are preferable. Examples of these are the spatial autocorrelation coefficients like the Moran's I (Reed and Burkhart 1985, Chou et al. 1990):

$$I = \frac{n\sum_{i=1}^{n}\sum_{j=1}^{n} a_i a_j W_{ij}(x_i - \bar{x})(x_j - \bar{x})}{S_0 \sum_{i=1}^{n}(x_i - \bar{x})^2} \tag{13}$$

where n is the number of stands, x is the value of the variable of interest, a is the stand surface area, and $S_0 = \Sigma\Sigma a_i a_j W_{ij}$ is the sum of the area-weighted spatial weights. In the simplest case, $W_{ij} = 1$ if stands i and j are adjacent, otherwise 0.

The spatial autocorrelation coefficient describes the smoothness of the change in the *HSI* or any other stand characteristics for which the correlation coefficient is computed. The only information needed for computing Moran's I is the adjacency information, i.e., the neighbour stands of all stands must be known. The disadvantage of spatial autocorrelation is that the coefficient must be recomputed after every change in the solution, using all stands, because a change in a single compartment may affect the mean value of the variable, to which the values of different stands are compared (see Equation 13). Because the effect cannot be localised, the rate of optimisation becomes slow.

The third spatial objective type tested by Kurttila et al. (2002) is the length of the boundary between neighbour compartments, both of which are suitable habitats (H–H boundary) or one of which is habitat but the other is not (H–NH boundary). Maximising the H–H boundary length tends to cluster and connect habitat patches, and maximising the H–NH boundary length tends to maximize the edge zone. The spatial information needed for calculating these variables are the lengths of boundaries that each compartment has with its neighbours. When evaluating the effect of a changed treatment of a compartment one needs to consider only this compartment and its immediate neighbours when calculating the move's effect on the total boundary length. Therefore, H–H or H–NH boundary length as an objective variable allows a reasonably fast optimisation. The drawback of this spatial objective type is that a clear limit must be specified for the *HSI* or the stand characteristic that determines whether a stand is a habitat patch or not. This drawback can be partially overcome by weighting the boundary section by the difference in the *HSI* between the two compartments. The calculation and the information contained in the lengths of boundary sections can be rather easily extended to cover core area and edge zone measures (Öhman and Eriksson 1998, Öhman 2000).

Although both spatial autocorrelation and boundary length can be used to either cluster or disperse resources, they may produce clearly different plans. The effect of a spatial objective type is illustrated in a planning example in which the habitats of flying squirrel were clustered or dispersed. Four 30-year plans with the following objectives were compiled for a forest area of 920 hectares:

- Plan 1: The length of habitat–habitat boundary was maximised
- Plan 2: The length of habitat–non-habitat boundary was maximised
- Plan 3: Spatial autocorrelation of *HSI* (Moran's *I*) was maximised.
- Plan 4: Spatial autocorrelation of *HSI* (Moran's *I*) was minimised.

The limiting value of the *HSI* for flying squirrel was 0.4 as in the study of Kurttila et al. (2002). Table 3 and Figure 9 show that the plans using these four variables as the sole objective are very different from each other. The mean *HSI* after 30 years is clearly the best when the length of H–H boundary is maximised. The connectivity of habitat patches is also good with this goal. Maximising spatial autocorrelation results in lower mean *HSI* than maximisation of H–NH boundary or minimisation of spatial autocorrelation. This is because continuous areas with low *HSI* values contribute to the Moran's *I* as much as continuous areas with high *HSI* values. Therefore, if spatial autocorrelation is used to cluster resources, another variable, like the total habitat area, should be used as an additional objective to guarantee that the optimisation algorithm does not maximise the autocorrelation by minimising habitat area.

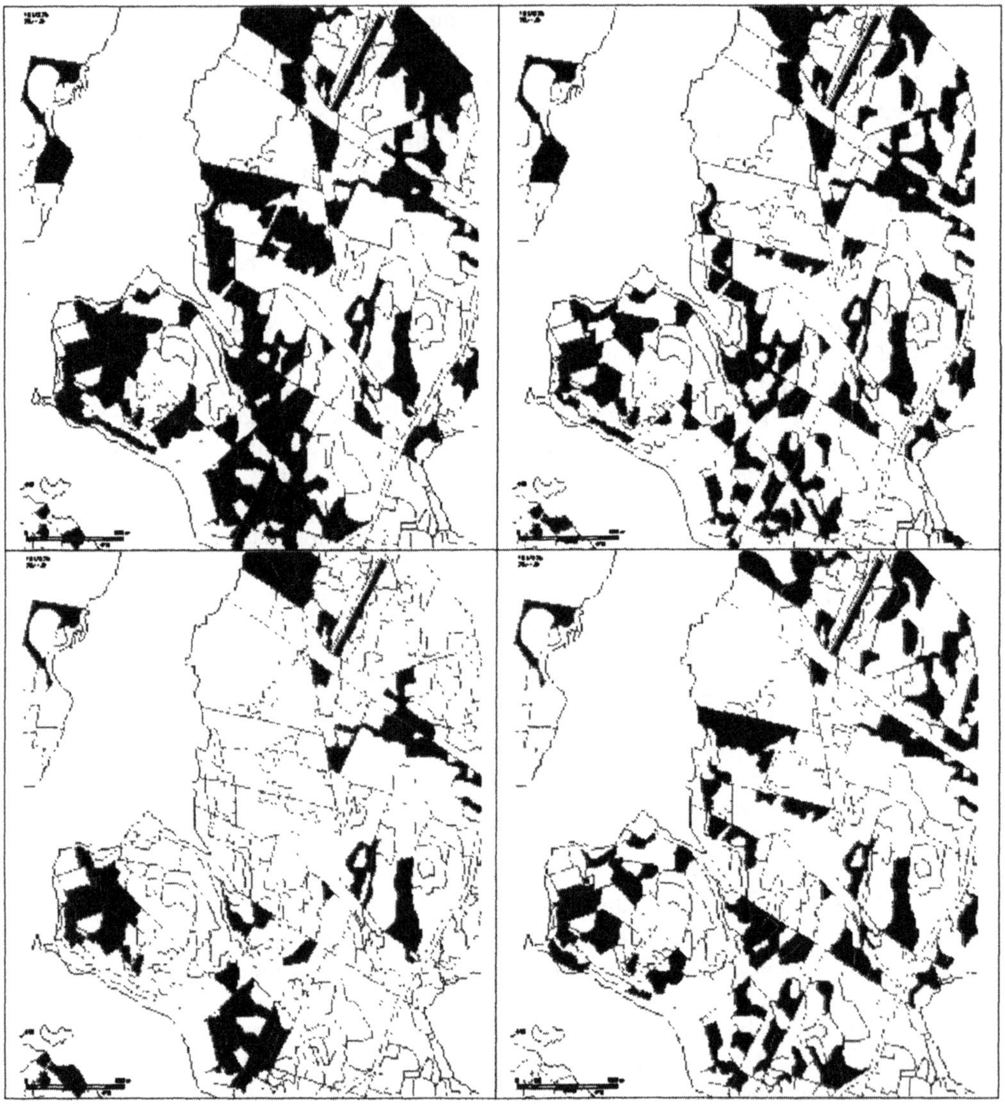

Figure 9. Value of the habitat suitability index (HSI) of flying squirrel in four forest plans (dark colour indicates high value; shades darker than ■ indicate HSI values greater than 0.4, and white colour indicates a value equal to zero). Top left: habitat–habitat boundary maximised; top right: habitat–non-habitat boundary maximised; bottom left: spatial autocorrelation of HSI maximised; bottom right: spatial autocorrelation of HSI minimised.

Table 3. The effect of objective variable (in single-objective optimisation) on the value of variables related to the ecological quality of a 920-hectare forest. HSI is the habitat suitability index of flying squirrel. H–H boundary is the proportion of habitat–habitat compartment boundary, and H–NH boundary the proportion of habitat–non-habitat compartment boundary.

	Objective			
Variable	*Max H–H boundary*	*Max H–NH boundary*	*Max spatial autocorrelation*	*Min spatial autocorrelation*
Mean *HSI*	0.327	0.294	0.209	0.215
Habitat proportion	0.243	0.231	0.200	0.225
Moran's *I* for *HSI*	0.218	0.009	0.461	-0.265
H–H boundary	0.182	0.062	0.091	0.029
H–NH boundary	0.311	0.509	0.232	0.399

The example shows that there are several reasonably simple alternatives to improve the spatial pattern of the forest landscape in numerical optimisation. However, different approaches proposed for the same purpose may result in distinctly different solutions. Therefore, it is important to know both the pattern of the landscape good for the species and the way different spatial objective variables behave in optimisation.

4. CONCLUDING REMARKS

This chapter reviewed some alternatives for numerically measuring the amount of forest outputs other than timber or economic profit. A common feature of the discussed methods is that they can be used in numerical optimisation, either as a component of the objective or penalty function, or as a constraint. However, this does not mean that numerical optimisation automatically produces a forest management plan that is optimal in terms of the non-wood outputs. There are several reasons for this. One is that numerical measurement of an output that is qualitative by nature is only a partial description of the forest quality. In addition, forest planning calculations always include a great deal of uncertainty, and this uncertainty may be much higher with non-wood forest outputs than with timber harvests. The solution also gives wrong suggestions if the optimisation problem is formulated in such a way that the objective function and the constraints do not reflect the true preferences of the decision makers.

Moreover, the decision maker may have difficulties in understanding what the abstract variables devised for non-wood outputs, such as stand's priority with respect to scenic beauty, or spatial autocorrelation of *HSI*, really mean. This problem is partly alleviated by calculating the numerical measures for the initial forest, which is familiar to the decision maker, as well as for the future states. In addition, it is useful to state the scale and range of variation of the variables used, so that the decision maker can specify reasonable target levels for the non-wood outputs. An interactive interface to the optimisation algorithm, with an opportunity to study the trade-offs between objective variables and the effects of changes in the objective function (e.g., Korhonen and Laakso 1986) further helps the decision maker to get a good grasp of the problem. The evaluation of alternatives may be further enhanced by instant computer graphics such as landscape visualisations (Pukkala and Pykäläinen 2000).

The solutions of forest planning problems should never be accepted without a critical evaluation of the results. The role of optimisation in forest planning is to produce decision alternatives (predictions, scenarios) rather than decisions. The alternatives should always be subjectively evaluated in subsequent analyses. If some of the models or model parameters are considered to be very uncertain, it is possible to study the sensitivity of the optimisation result to changes in this model or parameter. Another way is to produce several solutions, by modifying the problem formulation, and then use various comparison techniques to rank the alternatives. Several comparison techniques allow both qualitative and quantitative criteria to be included in the same analysis (see Chapter 3).

Some of the model types reviewed in this chapter predict subjective preferences rather than objectively measurable forest outputs. This means that the "forest model", which produces information for the planning model, is also subjective. However, this does not decrease the value of the numerical analysis because true decision making is always subjective and always depends on the preferences of the decision makers. What is important is that whenever subjectivity enters into the analysis the planning model and the predictions should reflect the preferences of the "right" persons, i.e., those who are the authorized decision makers for the planning situation at hand.

The many difficulties connected to the numerical measurement of non-wood forest outputs should not be used as an excuse for not using numerical models and numerical optimisation. Numerical predictions and solutions of optimisation problems usually tell part of the truth. The solution proposed by optimisation is efficient if the models and assumptions are correct. Efficiency means that pursuing a given management objective minimises losses in the other, usually conflicting goals. Taking into account the complexity of the planning problem in cases where there are multiple goals and decision makers and the area is not very small, it is not reasonable to assume that a planner or a decision maker can find efficient solutions without using efficient numerical tools. If the optimisation results are not trusted – they should never be fully trusted– they may be taken as starting points for more traditional development of forest management plans.

Another justification for using predictions of non-wood forest benefits in optimisation is related to the shape of the trade-off curves between timber production and the non-wood output. The curves are often very convex, which means that great gains in non-wood outputs can be achieved with only minimal losses in timber production goals (see Figs. 3 and 6). If the non-wood outputs are excluded from optimisation the solution may be slightly better from the timber production point of view, but much poorer for scenic beauty or biodiversity conservation. The production possibility boundaries also suggest that it is worthwhile to include the amenity score, biodiversity index, etc. in the optimisation problem even when these features are of low importance to the decision maker. With these features included, the optimisation results in plans which may be clearly better for society and not worse for the forest owner. In a large planning area there may be thousands of alternative plans which are almost equally good from the timber production standpoint but very unequal with respect to the non-wood outputs. Including the non-wood output in the optimisation problem, even with a minimal weight, is an easy way to pick the plan that maximises the non-wood forest output without seriously compromising the timber production targets.

REFERENCES

Alho J. M., Kolehmainen, O. and Leskinen, P. 2001. Regression methods for pairwise comparisons data. In: Schmoldt, D., L., Kangas, J., Mendoza, G., A. & Pesonen, M. (Eds.). *The Analytic Hierarchy Process in Natural Resources and Environmental Decision Making*. Kluwer Academic Publishers. Dordrech. Pp. 235-251.

Chou, Y., Minnich, R. A., Salazae, L. A., Power, J. D. and Dezzani, R. J. 1990. Spatial autocorrelation of wildfire distribution in the Idyllwild quadrangle, San Jacinto Mountain, California. *Photogrammetric Engineering & Remote Sensing* 56: 1507-1513.

Dykstra, D. 1984. *Mathematical programming for natural resource management*. McGraw-Hill, Inc. 318 pp.

Gadow, K. v. and Puumalainen, J. 2000. Scenario planning for sustainable forest management. In. Gadow, K. v., Pukkala, T. and Tomé, M. (Eds.). *Sustainable Forest Management*. Kluwer Academic Publishers. Dordrecht. Pp. 319-356.

Gustke, L.D. and Hodgson, R.W. 1980. Rate of travel along an interpretive trail. The effect of an environmental discontinuity. *Environment and Behaviour* 12: 53-63.

Hof, J. and Bevers, M. 1998. *Spatial* optimisation for managed ecosystems. Columbia University Press. New York. 258 pp.

Hytönen,L., Leskinen, P. and Store, R. 2002. A spatial approach to participatory planning in forestry decision making. *Scandinavian Journal of Forest Research* 17. 62-71

Ihalainen, M. and Pukkala, T. 2001. Modelling cowberry and bilberry yields from mineral soils and peatlands on the basis of visual estimation. *Silva Fennica* 35(3): 329-340.

Ihalainen, M., Alho, J., Kolehmainen, O. and Pukkala, T. 2002. Expert models for berry and cowberry yields in Finnish forests. *Forest Ecology and Management* 5413: 1-8.

Ihalainen, M. and Pukkala, T. 2001. Modelling cowberry (*Vaccinium vitis-idaea*) and bilberry (*Vaccinium myrtillus*) yields from mineral soils and peatlands on the basis of field estimates. *Silva Fennica* 35(3): 329-340.

Ihalainen, R. 1990. Structural changes in Finnish nonindustrial private forest ownership: A survey of the literature 1960-89. *Folia Forestalia* 750. 52 pp.

Ihalainen, R. 1992. Yksityismetsänomistuksen rakenne 1990. *The Finnish Forest Research Institute, Research Notes* 405. 41 pp.

Kangas, A. and Kangas, J. 1998. Ekologiset mallit ja ekologisten riskien hallinta metsäsuunnittelussa. *Folia Forestalia* 2/1998: 201-222.

Kangas, J., Karsikko, J., Laasonen, L. and Pukkala, T. 1993a. A Method for estimating habitat suitability on the basis of expertise. *Silva Fennica* 27(4): 259-268.

Kangas, J., Laasonen, L. and Pukkala, T. 1993b. A method for estimating forest landowner's landscape preferences. *Scandinavian Journal of Forest Research* 8: 408-417.

Kangas, J. and Pukkala, T. 1996. Operationalization of biological diversity as a decision objective in tactical forest planning. *Canadian Journal of Forest Research* 26(1): 103-111.

Kangas, J. and Mononen, A. 1997. Ekologiseen asiantuntemukseen perustuvan numeerisen mallin tuottaminen metsäalueen biodiversiteetin arviointiin. *Folia Forestalia* 2/1997: 225-238.

Kangas, J., Store, R., Leskinen, P. and Mehtätalo, L. 2000. Improving the quality of landscape ecological forest planning by utilizing advanced decision-support tools. *Forest Ecology and Management* 132:157-171.

Kangas, J., Pukkala, T. and Kangas, A. 2001. HERO: Heuristic optimisation for multi-criteria decision analysis. In: Schmoldt. D. L., Kangas, J., Mendoza, G. A. and Pesonen, M. (Eds.). *The Analytic Hierarchy Process in Natural Resource and Environmental Decision Making*. Kluwer Academic Publishers. Pp. 51-65.

Korhonen, P. and Laakso, J. 1986. A visual interactive method for solving multiple criteria problems. *European Journal of Operational Research* 24: 277-287.

Kouki, J. 1993. Luonnon monimuotoisuus valtion metsissä – katsaus ekologisiin tutkimustarpeisiin ja suojelin mahdollisuuksiin. *Metsähallituksen luonnonsuojelujulkaisuja* A11. 88 pp.

Kurttila, M. 2001. The spatial structure of forests in the optimization calculations of forest planning – a landscape ecological perspective. *Forest Ecology and Management* 142: 129-142.

Kurttila, M., Pukkala, T. and Loikkanen, J. 2002. The performance of alternative spatial objective types in forest planning calculations: a case for flying squirrel and moose. *Forest Ecology and Management* 166: 245-260.

McGarical, K. and Marks B. J. 1995. FRAGSTATS: *Spatial pattern analysis program for quantifying landscape structure*. USDA Forest Service, PNW, General Technical Report 351. 122 pp.

Mykrä, S. and Kurki, S. 1998. ESC-strategy for rational operationalisation of forest biodiversity maintenance in Finland. *Silva Fennica* 32: 389-399.

Nalli, A., Nuutinen, T. and Päivinen, R. 1996. Site-specific constraints in integrated forest planning. *Scandinavian Journal of Forest Research* 11: 85-96.

Naskali, A. 1991. Monikäytön ekonomia. In: Tapion Taskukirja. Kustannusosakeyhtiö Metsälehti. Helsinki. Pp. 271-285.

Næsset, E. 1997a. Geographical information systems in long-term forest management and planning with special reference to preservation of biological diversity. *Forest Ecology and Management* 93: 121-136.

Næsset, E. 1997b. A spatial decision support system for long-term forest management planning by means of linear programming and geographical information systems. *Scandinavian Journal of Forest Research* 12: 77-88.

Nousiainen, I. and Pukkala, T. 1992. Use of computer graphics for predicting the amenity of forest trails. *Silva Fennica* 26(4): 241-250.

Öhman, K. 2000. Creating contiguous areas of old forest in long term forest planning. *Canadian Journal of Forest Research* 30: 1817-1823.

Öhman, K. 2001. Forest planning with consideration to spatial relationships. *Acta Universitatis Agriculturae Sueciae, Silvestria* 198. 32 pp. + App.

Öhman, K. and Eriksson, L. O. 1998. The core area concept in forming contiguous areas for long term forest planning. *Canadian Journal of Forest Research* 28: 1032-1039.

Oksanen-Peltola, L. 1999. Metsäsuunnittelun lähtokohta. Metsäntutkimuslaitoksen tiedonantoja 741: 8-11.

Ovaskainen, V. and Kuuluvainen, J. (Eds.) 1994. Yksityismetsänomistuksen rakennemuutos ja metsien käyttö. *Metsäntutkimuslaitoksen tiedonantoja* 484. 144 pp.

Pukkala, T., Kangas, J., Kniivilä, M. and Tiainen, A-M. 1997. Integrating forest-level and compartment-level indices of species diversity with numerical forest planning. *Silva Fennica* 31(4): 417-429.

Pukkala, T., Nuutinen, T. and Kangas, J. 1995. Integrating the amenity of forest area into numerical forest planning. *Landscape and Urban Planning* 132: 185-195.

Pukkala, T., Kellomäki, S. and Mustonen, E. 1988. Prediction of the amenity of a tree stand. *Scandinavian Journal of Forest Research* 3: 533-544.

Pukkala, T. 1988. Methods to incorporate the amenity of landscape into forest management planning. *Silva Fennica* 22(2): 135-146.

Pukkala, T. and Pykäläinen, J. 2000. Super-HERO: Visuaalinen käyttöliittymä vuorovaikutteiseen optimointiin. *University of Joensuu, Faculty of Forestry, Research Notes* 105: 33-40.

Reed, D. D. and Burkhart, H. E. 1985. Spatial autocorrelation of individual tree characteristics in loblolly pine stands. *Forest Science* 31: 575-587.

Saaty, T. 1977. A scaling method for priorities in hierarchical structures. *Journal of Mathematical Psychology* 15: 234-281.

Saaty, T. 1980. *The Analytic Hierarchy Process*. McGraw-Hill. New York. 287 pp.

Shafer, E. and Brush, R. 1977. How to measure preferences for photographs of natural landscapes. *Landscape Planning* 4: 237-256.

Siitonen, M. 1999. Metsien käsittely endogeenisesta analyysista – näkemyksiä ja näkökulmia. *The Finnish Forest Research Institute, Research Notes* 725, Pp 10-26. [In Finnish]

Silvennoinen, H., Alho, J., Kolehmainen, O. and Pukkala, T. 2001. Prediction models of landscape preferences at the forest stand level. *Landscape and Urban Planning* 56(1-2): 11-20.

Store, R. 1996. Maiseman huomioonottavan metsikkökuvioinnin tuottaminen paikkatietojärjestelmällä. *Folia Forestalia*. 1996(3):245-262.

Store, R. and Kangas, J. 2001. Integrating spatial multi-criteria evaluation and expert knowledge for GIS-based habitat suitability modelling. *Landscape and Urban Planning* 55:79-83.

Tyrväinen, L. 1999. Monetary valuation of urban forest amenities in Finland. *Finnish Forest Research Institute, Research Papers* 739. 55 pp + App.

GPSR Compliance

The European Union's (EU) General Product Safety Regulation (GPSR) is a set of rules that requires consumer products to be safe and our obligations to ensure this.

If you have any concerns about our products, you can contact us on ProductSafety@springernature.com

In case Publisher is established outside the EU, the EU authorized representative is:

Springer Nature Customer Service Center GmbH
Europaplatz 3
69115 Heidelberg, Germany

Batch number: 10873406

Printed by Printforce, the Netherlands